T0317528

STEPS TO
SAFETY CULTURE
EXCELLENCE^{SM}

STEPS TO SAFETY CULTURE EXCELLENCE^{SM}

TERRY L. MATHIS
Chief Executive Officer
ProAct Safety, Inc.

SHAWN M. GALLOWAY
President and Chief Operating Officer
ProAct Safety, Inc.

A JOHN WILEY & SONS, INC., PUBLICATION

Copyright © 2013 by John Wiley & Sons, Inc. All rights reserved

Published by John Wiley & Sons, Inc., Hoboken, New Jersey
Published simultaneously in Canada

No part of this publication may be reproduced, stored in a retrieval system, or transmitted in any form
or by any means, electronic, mechanical, photocopying, recording, scanning, or otherwise, except as
permitted under Section 107 or 108 of the 1976 United States Copyright Act, without either the prior
written permission of the Publisher, or authorization through payment of the appropriate per-copy fee
to the Copyright Clearance Center, Inc., 222 Rosewood Drive, Danvers, MA 01923, (978) 750-8400,
fax (978) 750-4470, or on the web at www.copyright.com. Requests to the Publisher for permission
should be addressed to the Permissions Department, John Wiley & Sons, Inc., 111 River Street,
Hoboken, NJ 07030, (201) 748-6011, fax (201) 748-6008, or online at http://www.wiley.com/go/
permissions.

Limit of Liability/Disclaimer of Warranty: While the publisher and author have used their best efforts
in preparing this book, they make no representations or warranties with respect to the accuracy or
completeness of the contents of this book and specifically disclaim any implied warranties of
merchantability or fitness for a particular purpose. No warranty may be created or extended by sales
representatives or written sales materials. The advice and strategies contained herein may not be
suitable for your situation. You should consult with a professional where appropriate. Neither the
publisher nor author shall be liable for any loss of profit or any other commercial damages, including
but not limited to special, incidental, consequential, or other damages.

For general information on our other products and services or for technical support, please contact our
Customer Care Department within the United States at (800) 762-2974, outside the United States at
(317) 572-3993 or fax (317) 572-4002.

Wiley also publishes its books in a variety of electronic formats. Some content that appears in print
may not be available in electronic formats. For more information about Wiley products, visit our web
site at www.wiley.com.

Library of Congress Cataloging-in-Publication Data:
Mathis, Terry L.
  STEPS to Safety Culture Excellence[SM] / Terry L. Mathis, Shawn M. Galloway.
    p. ; cm.
  Includes bibliographical references and index.
  ISBN 978-1-118-09848-6 (cloth)
  I. Galloway, Shawn M., 1976– II. Title.
  [DNLM: 1. Occupational Injuries–prevention & control. 2. Safety
Management–organization & administration. 3. Accidents, Occupational–prevention &
control. 4. Occupational Health Services–organization & administration. 5. Organizational
Culture. 6. Organizational Innovation. WA 485]
  616.9'803–dc23
                                   2012035830

10 9 8 7 6 5 4 3 2 1

CONTENTS

* Safety Culture Excellence is a trademark of ProAct Safety, Inc. and is registered as U.S. Service Mark Registration No. 3,972,134.

INTRODUCTION

There is no substitute for excellence, not even success.
—*Thomas Boswell*

There are two kinds of people in safety: the kind who care and the kind who do not care. Those who do not care do not work toward excellence because they do not care! The people who do care are the kind who change the world for the better and the ones we are proud to work with and to help. For these people, nothing short of excellence is "good enough" in safety. Whether you are a safety professional, a concerned manager, a union safety representative or the president or chief executive officer (CEO) of the organization, we would like you to consider what safety excellence could mean for you and your organization.

What is the public image of your organization now and what will be the legacy you leave behind? How would you like it to be known and remembered? Have you ever thought or dreamed that you would like to be part of the organization that cured cancer or heart disease? How about being part of an organization that conquered a bigger killer than either of those terrible diseases? We are talking about accidental injury! Do you realize that this terrible and preventable tragedy takes the lives of more people between the ages of 1 and 44 than either of these diseases? It is among the top 10 causes of death in every age group.

When you help your organization develop Safety Culture Excellence^{SM}. you improve the quality of life for everyone who works there. You help them and their families to avoid not only the deaths but also the debilitating, expensive, and lifestyle-destroying injuries that can result from on-the-job accidents. If you truly change the safety culture at work, you are likely to impact off-the-job injuries as well. You will give the people you work with the gift of an accident-free life and the skills to duplicate it year after year. You will give them the structure and capabilities to attack safety challenges one at a time and to conquer them. That structure and capability will help you address virtually every process and significant element of business organization that impacts safety and will make them foster and reinforce excellence.

Development of Safety Culture Excellence is altruistically rewarding and not bad for the business bottom line either. You will find yourself among other organizations that have created safety excellence and expect it of their associates, clients, and suppliers. You will find organizations further along the path who are glad to help and those behind you eager for your assistance. You will find yourself among not only those with like minds but also those with hearts deeply committed to helping people through the pursuit of safety excellence.

You will find that your culture perpetuates excellence and that its excellence in safety tends to grow into excellence in every other aspect of your organizational operations. Excellence produces pride, and pride produces even more excellence. This is not a poet's dream or an empty promise from someone with something to sell; it is a reality that has already been accomplished by a number of organizations. Many others are beginning the journey.

Always remember that excellence is not necessarily perfection; it is more like personal best. Can your organization be its best and expand its capabilities beyond what it once thought possible? We believe it can. If you believe it also, come join the journey beyond bad, beyond good, beyond great to the highest level of performance possible within your organizational realities. Achieve excellence in safety and align your culture to ensure that it is sustainable into the bright future you will create.

The very idea of improving a whole culture of hundreds or thousands of people can seem daunting, but it has been done many times successfully and can be broken down into bite-sized pieces we call STEPS. When you begin to work on your safety culture a STEP at a time, you create momentum. You instill into your culture the seeds of excellence. A culture that can take a single STEP toward improved performance can take another, and another. Every journey, no matter how long, is made up of STEPS. Learning to STEP is learning how to improve. Learning how to improve is developing the basic skill of excellence.

Even though we focus on the safety aspect of excellence, the process we are proposing can be used to create excellence in any aspect of organizational performance. There are advantages to starting with safety. Safety is altruistic and tends to get the hearts and minds into the effort rather than simply hands and feet. It boldly answers the what's in it for me (WIIFM) question. It benefits every employee, their families, the community, and the organization. Once it becomes a success in safety, it can be turned toward other targets and produce a wealth of organizational excellence.

The journey to Safety Culture Excellence will take you through a series of STEPS designed to help you reach seven milestones. Each milestone is an aspect of cultural excellence. You may find that you have already taken some of these steps and can reach a milestone quickly with less effort. Some STEPS may need to be revisited in years to come. The STEPS leading to the first five milestones are designed to create a culture of excellence in which continuous improvement is not only possible but also reinforced and empowered at every level. The STEPS leading to the sixth milestone are designed to create the capability within the culture to identify, prioritize, and solve safety problems and challenges. The seventh milestone contains STEPS to maintain and continuously improve the excellent performance of the safety culture.

Case Study: We were working with an organization that had multiple sites with varying safety performance. However, one site had a perfect safety record for over 15 years and was the only site without a safety professional on staff. We asked to visit the site and acquired the proper personal protective equipment (PPE) and visitor identification. When we drove into the parking lot near the front office, a worker in a company truck saw me approaching and parked next to us. He introduced himself and examined our credentials and offered to accompany us on our visit. We checked in to the security office and proceeded to tour the facility. He told us of a well-respected safety professional who had established their safety programs many years ago and had then retired. We found out that our guide was not assigned, but that virtually anyone who saw a visitor approaching would have taken the same initiative. During our visit, everyone we observed was looking out for each other and offering safety information to us for each area we entered. The site had a relatively stable population with low turnover, but more notably, it had a safety culture in which everyone was focused and involved. It was the culture that was producing the excellent safety performance, even without an official safety leader. We studied it closely to help the organization adjust the cultures at the other sites and develop some of the same capabilities.

VISION

Excellence is a journey, not a destination. Those who think they have reached excellence and stop their travels find that their goal has evaded them. The perfect ending of every journey is not where it takes you, but what it makes of you. This journey to Safety Culture Excellence is ongoing and enhances your capabilities with each STEP.

We, the authors, believe that safety is both the ultimate humanitarian cause and the most valuable of strategic advantages for organizations. Those who are best in safety will attract not only the finest talent, but the most wonderful human beings. They will have workplaces that foster creativity and job satisfaction. They will be appealing partners to firms that need their products and services and will win the richest contracts. Those who are best in safety will be willing to share their safety successes with their business partners, their employees' families and their communities. Safe organizations care about people and that caring does not stop when people go out the front gate.

The STEPS process will demystify safety. It will no longer seem impossible, vague, overloading, or evasive. An organization can determine a starting place and develop a map to success. The journey can be self-paced and will suit itself to the inevitable variables between cultures. The goal is not perfection, but personal best. Each culture can begin a journey toward its own ultimate level of excellence. No one will be required to make radical, sudden changes to their styles or practices, but rather gradual evolution toward a more perfect and harmonious way of working together and sharing the joy of accomplishment.

If this vision sounds idealistic, please remember that we who created it are among the most practical and successful safety excellence consultants in the world and that we are the staunchest critics of theories that cannot work in the realities of today's workplace. This approach is based on research, but also on sound principles learned in the real world with real successes. The ultimate research is carried out in the laboratory of human endeavor and the kind of peer review we seek is the success of those who utilize our ideas and methods.

UNDERSTANDING THE CONCEPT OF SAFETY CULTURE

"Safety culture" is a term in much use today in the safety community. Organizations are realizing that top-down programs and management tactics have limitations when they meet the real-world workplace. Culture was mentioned as an underlying cause in the most noteworthy recent disasters. Organizations such as NASA and BP were accused of not having a good safety culture that could prevent disasters such as Challenger, Texas City, and the Gulf Oil Spill.

The notoriety of safety culture has caused many organizations to question their own efforts. Are they doing what it takes to create and encourage the right kind of culture to avoid the costs and negative publicity of similar disasters? The number of articles written on safety culture has grown exponentially, as have the hits on related websites. Our own clients and prospective clients are constantly asking about culture. Even while pursuing other programs or processes, they are concerned how these will impact their safety culture.

No one wants to be ambushed with disaster and bad publicity, and it seems that nothing can produce ambushes so well as the unpredictability of safety issues. Most organizations put a lot of effort into safety, which can multiply the disappointment when things go wrong. So, how do we take the uncertainty and guesswork out of safety? Alan Kay of Apple computers said, "The best way to predict the future is to create it." Developing a strong and capable safety culture is the best way to control (by creating) your own safety future.

However, unlike others in this field, we are not advocating "creating" a safety culture from scratch. You already have one! What we are suggesting is that the culture can be better and that a better culture will build sustainability into your safety efforts more effectively than any other approach known to date. A culture is what is shared among the members of an organization. Those shared events and perceptions influence personal and organizational performance, and can either encourage or discourage the growth of excellence. To shape your existing safety culture in a way in which excellence can and does grow is the goal of this book.

The title of this book contains several terms that should be defined clearly before moving on to the "how-to" parts. Clearly defining the crucial terms can aid greatly in making sure that everyone is on the same page, philosophically and strategically, before beginning the process. Clearly defined terms create common vision. One of the goals of this book is to achieve the type of clarity and alignment that excellent safety cultures possess. So, starting with the same vision is especially important to achieving this goal.

Before we address the terminology in the title, let us define and discuss a term implied in every safety effort: "accident." If this word has negative connotations to you and you would like to use another term in your safety efforts, please do so. However, in this text, we will use the term and define it in this way: An accident is either something that is done unintentionally or something that is done deliberately that results in an unintended outcome. In short, accidental is the opposite of intentional or deliberate. This term is going to be used as an adjective before the term "injury," which is already well defined. Accidental injuries are injuries that were neither intentional nor deliberate and can vary in severity from minor to catastrophic. We will later discuss this and other definitions and suggest that you either adopt these or create your own. Definition is an essential part of clarity that helps lead to excellence.

The first term in the book title is *STEPS*. Each letter is capitalized because STEPS is an acronym as well as a word. It stands for Strategic Targets for Excellent Performance in Safety[SM]. So, the term STEPS is intended to connote both the steps toward an ultimate goal and as well as the process to choose and execute each of the steps. One of the main sources of failure in safety-improvement efforts is trying to do too much at once. Working on unclear goals or focusing on the wrong things is the next most common. For these three reasons, it is critical to take the right steps, the right-sized steps, and to take them in the right order.

The next term to define is *safety*. Almost everyone will tell you that they know what safety is, but when asked to define it, most will give you the goal rather than the definition. The most common responses are that safety is "not getting hurt," or safety is "going home exactly the way you came to work (with all your body parts intact)." Obviously, these are the goals of safety, but what is the definition? If a group of people is going to work on a goal, the goal must be clearly defined and universally shared. So, let us begin with a very generic, 30,000-ft definition of safety: safety is knowing what can hurt you, learning the things that can keep them from hurting you, and doing those things.

We encourage you to create your own definition of safety, but we want to start here with a basic meaning to help you understand what success looks like. This is, of course, an oversimplification of all the ways we work in safety. It does not elaborate on the methodologies of mitigating risks, but it focuses us on the basic objectives. Obviously, if we are to anticipate and avoid injury, we must identify the risks and address them. Almost all accidents can be categorized as a failure to identify risks or a failure to adequately address the risks. We will use this dichotomy in our methods to identify the best way to prevent accidents. Asking the question, "Was the accident the result of an unidentified or underestimated risk, or was it a failure to adequately address a known risk?" can help to determine the best prevention strategy. Even though the difference between these two might seem slight, the methods used to address them can be radically different. Also, in our consulting practice, we have identified that the failure to recognize this difference often results in using the wrong solutions to solve safety problems.

There is a school of thought applied to safety called Human Performance Improvement (HPI). While HPI tends to focus on what they call "human error" and to classify the types of error, they readily admit that people get hurt for three basic reasons: they fail to recognize the risk, they fail to take a precaution, or the precaution they take is inadequate to address the risk. We would add to the first category that often the failure is not the failure to recognize the existence of the risk, but it is an underestimating of the probability of that risk to result in an accidental injury.

Taking this into consideration, one could develop a similar definition of safety: recognizing risks and taking adequate precautions to prevent the risk from resulting in injury.

There are two other concepts that should be mentioned in any discussion of safety: the idea of conditional safety and the idea that human behavior is often influenced by systems issues beyond the control of the workers involved. Safety cultures must, to whatever degree possible and reasonable, address unsafe conditions. We believe that some programs for workplace safety put unreasonable hopes on conditional fixes and that it is impossible to remove all hazards. That being said, almost all workplaces have additional opportunities to improve safety conditions. As for the second argument about behavioral causation, we have always recognized that human behavior cannot be the root cause of accidents, simply because there is a reason for the behavior (another "why" in the causal chain). However, it is wise to recognize that worker behaviors directly impact safety outcomes, and that identifying crucial behaviors (such as precautions) and controlling the systems and other issues that influence them are valid methods to improve safety outcomes. So, a complete approach to safety must include workplace conditions and common work practice. STEPS includes methods to address both and to formulate improvement tactics that combine the two.

There has been a lot of discussion of late on the term "behavior." There is an effort to challenge, at least, the well-accepted concepts that "unsafe acts" cause the majority of accidents and to what degree workers have control over these acts. In Europe, this discussion has spawned such terms as "multiple causation" or the idea that accidents can be caused, impacted, or influenced by multiple factors. Even if some of the terminology is nonstandard, these basic ideas are not new. The fact that they were not en vogue or widely accepted does not negate the fact that they have been often challenged. In fact, it is the addressing of these potential factors that influence human behavior and how we handle conditional issues that has given rise to the thinking underlying the approach recommended in this book. We cannot ignore issues that influence risks, even when they do not result in accidents. We cannot ignore "unsafe acts" even when they are beyond the control of the individual. It is important to remember that our goal is to prevent, not to analyze. Even accidents that are not "caused" by behaviors can sometimes be prevented by them. Likewise, we cannot ignore unsafe conditions even when we do not currently have the knowledge or resources to eliminate them. The STEPS methodology suggests that we systematically and progressively assess and address each of these issues and ensure that no element of our safety culture is reinforcing risks.

The third term is *culture*. It is this word that prompted the writing of this book. There is no shortage of books with the term "safety culture" in their title or subtitle. But after reading them all, we feel that we have significantly more to offer. Each of the books we read addressed the subject, but none of them, in our professional opinions, practically and holistically mapped out a path to help an organization achieve and sustain safety excellence. There is a lot of academic work on organizational culture that has been almost totally neglected by the academics writing about safety culture. One excuse for this diversion is that, theoretically, there is no such thing as a safety culture. Cultures have been traditionally defined as commonalities of a group of people, not limited to a particular subject or goal. But almost everyone agrees that cultures impact safety and that improving the safety aspects of an organizational culture can greatly improve safety performance. This approach is what we refer to as Safety Culture Excellence.

A critical element to improving a safety culture involves changing a basic paradigm of what a safety culture is. Is a safety culture something an organization *has*, what the organization *is*, what the organization *does*, or *why the organization is the way it is*? Is it a state of being or a dynamic feature of performance? Is it passive or active? Most definitions of safety culture define the culture's characteristics. The definition we propose involves developing a culture's capabilities. The most basic capability of a safety culture is the ability to improve. That is the real challenge; not what managers or consultants can make of a safety culture, but what a safety culture can make of itself. Once a culture can take a STEP toward better performance, it can continue to take STEPS until it achieves its personal best. So, the real question is not, "What is our safety culture like?" but rather, "What can our safety culture do?" How do the norms of the group influence individuals within the group when they make safety decisions or follow common practice? Can the group learn to improve its own norms, common practice, and the ways in which it influences its members?

To use this active rather than passive definition of culture is a step away from the traditional notion that leadership is the primary element of culture. The underlying theme of many of the safety culture books is still the archaic notion that the workers should "do as they are told," and there are some tricks to make them more docile. In our extensive experience with almost 2000 sites in over 40 countries, we have yet to see a do-as-you-are-told culture reach safety excellence. John Reinecke and William Schoell in their *Introduction to Business* said, "Leadership is a manager's ability to get subordinates to develop their capabilities by inspiring them to achieve." We believe that this type of leadership is the way to create a culture that grows in capabilities and thrives on achievement without being told to do so. These are the types of cultures that we have seen create safety excellence.

This is not to say that leadership has no role in the safety culture other than to inspire. In fact, they play a critical part in developing the culture and in establishing excellence as a goal. Without the right leadership and reinforcement, it is virtually impossible for any group of people to develop the capabilities and use them to achieve excellent performance. But the role of leadership cannot and should not be defined within the antiquated theories of command and control. The role of leaders is to set levels of expectations, provide resources, empower, and allow the culture to continuously improve. Leaders can expect excellence, but they cannot demand it. They can help it happen but cannot accomplish it by decree. Helping leaders understand their roles accurately and fulfill them systematically is absolutely necessary for Safety Culture Excellence.

An often missed or understated aspect of culture is that, once established, it tends to perpetuate itself for generations. Therefore, we want to reassert our position that culture is a sustainability tool. It will impact safety not only in the here and now but also far into the future. So, an investment in developing a safety culture can pay big dividends for years, if not decades. Culture, when developed at the worker level, also tends to outlive changes in management, ownership, mergers, policies, laws, and other influences. When a culture develops safety practices, those practices become a norm that adapts and perpetuates in amazing ways.

So, the Safety Culture Excellence we are working toward is not a destination where everyone can rest and pat themselves on the back. It is a set of capabilities that enable continuous improvement in safety performance and create a chemistry and climate in which such improvement is nurtured and encouraged. In other words, it is a journey toward excellence, a journey with intermediate milestones along the way, but no final destination other than the ability to continually take steps toward greater excellence.

The final term is *excellence*. For our purposes, excellence is not necessarily perfection. Far too many academics and consultants tend to define safety perfection and challenge organizations to adapt their culture to a perfect model. Unfortunately, this seldom happens. Imperfect cultures do not completely remake themselves following a model of perfection. Excellence in safety is more akin to "personal best" than to perfection. Whatever the current level of safety performance in an organization, there is almost always a potential to be significantly better. Often, striving for perfection undermines the ability to become better. The perfect gets in the way of the good. When you try to take the whole safety excellence journey in one step, you almost always fail. When you take it a step at a time, choose the right steps in the right order, and develop a culture of always becoming better, you will almost surely succeed. Excellence is a journey toward perfection with the realization that perfection is a moving target, and there will probably always be another step between the organization and the ultimate goal.

So, the kind of culture we are aiming at is not academically perfect, ideal, or having every desired characteristic. We are aiming at a culture that can clearly understand its current state and target stepped improvements. This capability empowers a culture to improve its safety performance to a level of excellence that is not just sustainable but can be continuously improved. This culture will not be dependent on new programs or processes, but it will have an improvement methodology and mindset woven into the very fabric of the common practices and the addressing of workplace conditions that impact safety. As you begin the journey, you will find that each STEP helps you develop a capability. Each of these capabilities plays a crucial role in helping you to know where you are on the journey, maintain clarity of purpose, and have the kind of climate and chemistry necessary to address your risks and grow your safety culture to excellence.

We, the authors, think that there is a tendency to oversimplify safety into a basic, linear, and cause-and-effect model. We need to think more in a causal chain mentality and to create a balanced scorecard in which we recognize how much effort produces a change in perception, which creates a change in behavior, and which impacts accident experience by a certain reduction in frequency or severity. Simply trying to eliminate accident "causes" reactively has led to limited success. In a STEPS process, you can address and align all the major causes, contributing factors, influences, barriers, obstacles, and other factors that impact safety performance. By systematically looking at each of these, you can begin to gain a true insight into how safety excellence can be achieved.

We realize that there are endless philosophical points of view and countless research projects that could be used to argue that other criteria are more important than the ones we choose. However, we are suggesting that a course to excellence should include visiting each of these areas and *then* developing a methodology to address any and all other factors. Such an approach might have an imperfect beginning but will lead to a more perfect final destination.

Excellence is a journey, not a destination, an event, or state of being. It is best measured by progress, not status. The only status that should concern an excellence-based culture is direction and velocity. Like a great piece of music, excellence is something to be mastered and practiced, not simply played with a minimum of error.

We strongly recommend that you resist any urge to read only a chapter or two and jump into a safety-improvement initiative based on this methodology. The path to Safety Culture Excellence described in this book is dependent on constantly keeping the big picture or roadmap in mind while working on any given STEP. This means that you should read the entire book and make sure that you understand thoroughly how to proceed and what will be required before you begin. There are also some redundant sections in which we first ask you to consider certain elements while formulating a strategy and then come back to the same elements and ask you to address each one as a STEP to a more perfect safety culture. Reading the more complete information in the later section will help you better address each element in your strategy development.

Even if you are already on a path to Safety Culture Excellence and intend to "cherry pick" this book for ideas to enhance your efforts, we believe that you can do that best by completing the book before trying to implement any particular aspect. Even if you are only going to use a single idea, you will understand that idea more fully after a complete read.

We also believe that answers are important but that the right questions are even more important. In many sections of this book, we will ask questions, the answers to which will vary from one organization to another. It is not our goal to have every answer. It is our goal to help each of you to ask the right questions to learn the best way to achieve Safety Culture Excellence. In many of our projects, our clients have told us that the questions we asked helped them to self-discover the best ways to move ahead. We have tried to include many such questions in this book in the hopes that you will have a similar experience and make valuable discoveries with your own knowledge of your own organization. Put on your thinking cap!

OTHER WORKS ON
SAFETY CULTURE

After reading our first baker's dozen books on safety culture, we found some themes on which we will comment:

- The term "safety culture" is useful, nifty, and ought to continue, although it is probably an oxymoron and definitely not discretely defined. (We agree.)
- Safety culture can and should be "managed" and that it all must start with the leaders of the organization. (We disagree; leadership is important, but management is only one tool in shaping culture and not the only starting point.)
- The way to measure safety culture is through a perception survey and that a cumulative total of perceptions approximately describes the culture. (We disagree; it is one measure of one element of culture, but it is far from a complete, discrete, or accurate metric.)
- "Changing or creating" a safety culture will take years and a lot of patience, not to mention consulting fees. (We obviously strongly disagree, although the fee part does not sound that bad.)
- Certain programs such as Lean Six Sigma or behavior-based safety (BBS) can be effective tools to improve safety cultures. (We agree that such programs can impact safety culture but do not consider them sufficient alone to address all the aspects of Safety Culture Excellence^{SM}.)

While each of the books we read had something of value, they suggested to us that the thinking and methodologies to create Safety Culture Excellence are far from mature. The approaches recommended to improve safety cultures vary greatly and suggest that we are still searching for the best practices and terminology to empower our efforts and to accomplish our goals.

Several of the academics questioned why the research on organizational culture, of which there is a great deal, was not used as the foundation for studying safety culture. The problem with that was described by Frost, one of the academic researchers, as follows:

> Organizational culture researchers do not agree about what culture is or why it should be studied. They do not study the same phenomena. They do not approach the phenomena they do study from the same theoretical, epistemological or methodological points of view. (Frost et al. 1991)

In other words, there is no real agreement on organizational culture even after many years of study. Little wonder that this research was not used as a foundation for studying safety culture. It is also not surprising that after only a few years of discussion, we are not in complete agreement about safety culture.

Several of the more practice-based books tend to describe safety culture based on a reduction model of best practices among the organizations that reported the most zero-recordable or zero-lost-time records. They tout the virtues of setting zero as an accident goal or "safety first" as a mantra to rally the culture. They also tout the critical role of management and leadership in steering the organization toward those goals. They include stories of organizations that used particular programs or management practices and achieved excellent safety results. The theme seems to be that imitation of successful organizations can lead to success.

We seriously question imitation as a tool for Safety Culture Excellence. Just as copying successful individuals is not a guarantee of personal success; copying an organization that produces successful safety results is not a guarantee that you will achieve similar results. Even creating similar results does not guarantee that you will create a culture that can repeat those results. Cultures are so diverse that trying to get one to imitate another is somewhere between impractical and impossible. Also, cultures that produce excellent results in safety lagging indicators do not always accomplish it the same way. What works for one does not necessarily work for another. Also, many organizations accomplish great results in the short term but cannot sustain them. That is no model to imitate. Others are successful in spite of certain practices. Copying those is no guarantee of success either.

While we found something valuable in all these works, we were reminded of the parable of the blind men and the elephant. Each grasped a different part of the elephant (an ear, a leg, the trunk, and the tail), described it accurately, and developed logical conclusions and associations (an elephant is like a fan, a tree, a hose, a rope, etc.). The trouble was in subsequently ascribing the characteristics of the part to the whole. Safety culture is a rich topic, and many interpretations are possible. We viewed these works as sincere efforts to achieve a very complex and worthwhile goal. Many people are exploring to find the best path to Safety Culture Excellence, and many different trails are being followed in that exploration. We are writing this book, not to discredit those but to build on what they have started in an effort to take that exploration to the next level.

What we are recommending is a reasonably complex view of culture with specific, but customizable, steps to reach the ultimate goal. Not only is each step customizable but it can also be skipped or substituted to meet specific needs. This approach requires a savvier and more sophisticated implementation, but the methodology is also more sound. In contrast to an approach that simply allows you to study and appreciate your own complexity or an approach that specifies each inflexible step, this approach requires some tailoring and fitting. That means the implementer needs to be able to think, not just to follow instructions.

Although the journey to Safety Culture Excellence can take various courses, there is a roadmap and a good set of travel guides for whichever route you take. The goal is to remain flexible to the site's and/or organization's specific issues while remaining true to the guiding principles. This is not an exploratory journey into the complete unknown nor does it require advanced degrees or superhuman skills. We have trained literally hundreds of internal consultants in organizations large and small, and they have been successful. But before we can begin the journey, we need to define the goal of this trip and the rationale for taking it.

UNDERSTANDING AND IMPROVING SAFETY CULTURE

Since every culture is different, every starting point on the journey to Safety Culture Excellence^{SM} will be unique. In Milestone 2, we will recommend you to do an assessment of your starting point. For now, we would like you to consider some factors that will impact the way you move forward. This section is designed to help you understand better what is involved in improving a safety culture and to encourage you to consider some variables that can impact your path. Consider these aspects of safety culture and begin to analyze the starting point of your workforce teamwork, supervisory style, management and leadership style, and aspects of your workplace and workstation design and programs that impact culture.

It is important to remember that the purpose of this trip is to benefit the travelers and not to reach a specific destination. The journey to Safety Culture Excellence is a bit like a pilgrimage that we take for deeper purposes than travel. It is a game we play not only to win but also to build teamwork. We will create experiences in this journey that will test our determination and expand our capabilities. To get the most out of the journey, the travelers should know the rationale, the reasons we think it is important, and what we hope to accomplish. They should also come to know their fellow travelers and what role they will play in the journey. Consider the following issues as you begin to plan the trip.

We have attempted to at least begin a definition of safety culture, but ours is not the only one. Although many writers and experts are talking about safety culture, there is no real consensus on what it is. Almost everyone has a mental image, a set of ideas, and a particular example that comes to mind when they hear the phrase. But these images and ideas are not the same and do not necessarily match what others are thinking. Some view culture as something an organization "has," and others see it as what the organization "is." Some view culture as something that can be designed and dictated by management, and others see it almost completely in the hands of the workforce. Some describe culture as something that goes on in the heads of workers, and others think that it is more what goes on between the heads of the workers and defines them as a group. It is good to have a diversity of ideas on the subject, but if we are going to understand and improve this thing we call a safety culture, we need to narrow down these ideas and better focus on the work at hand. How do your employees think of safety culture? Are their concepts similar or vastly different?

To describe or compare and contrast the academic definitions of safety culture is both beyond our scope and nonproductive to our purpose, but certain aspects of them are relevant. The scholarly definitions of culture tend to center around shared or common values and beliefs and move on to the ways in which these become communicated and shared among the group forming the culture. Once shared, these values and beliefs manifest themselves as common practice. The scholars are beginning to expand their definitions beyond the individuals forming the group and their interrelationships to include the environment and context in which these individuals operate. In the past, there has been a tendency to try to isolate the culture from its environment to study it. Such isolation is a part of what is called "scientific method" and is designed to understand the studied part without the complications of other parts.

The problem with this approach to studying culture is that the two, culture and environment, lose their context when separated. When you separate these two, you tend to assume that things that are not true. For example, the people in a television reality show do not share the same environment as soldiers in a war zone and therefore form different cultures even when performing similar tasks. If you ignore the environment, you tend to assume that the commonality of tasks would be the main impact on the cultures. The movement toward more ethnographically inspired methods of looking at culture have greatly improved the academic approaches to understanding culture, although they have not resulted in any real consensus on what culture is among the academics using the methods.

Most of the nonacademic definitions of culture we have encountered are more definitions of common practice than culture. Phrases like "the way we do things around here" and "what people do when no one is looking" are more definitions of common practice than of culture. Culture becomes visible when it turns into practice just as attitude becomes visible when it turns into behavior. But the practice is not the culture just as the behavior is not the attitude. Culture is deeper than practice. It includes the reasons for the practice, the influences on the practice.

Case in Point: There are often stories in psychology or behavioral science textbooks illustrating how common practices occur and endure: stories of women who baked a roast in two pans because their mother always did it that way and found that grandma started the practice because she did not have a pan large enough for the whole roast, or stories of young engineers who had to design equipment to fit on rail cars and found that the reason for the distance between rails went all the way back to size of the rear end of a Roman horse. These stories illustrate how cultures tend to perpetuate practices even when the reason for doing so is removed and forgotten. They also show how long such practices can be perpetuated in the culture and how that could potentially impact safety practices that become cultural norms.

To understand why people do things in a certain way, we need to examine these reasons and influences and find which of them are particular to this group of people. It is crucially important to remember that culture is not just the sum of the individuals; it is the points at which the individuals connect and agree. Deming said that if you take workers out of one plant and put them into another, they will behave differently. You have not changed the worker, you simply have put the worker into a different culture, and the culture will have new influences. So, what and who is it in a culture that influence the workers?

The first, obvious answer is "each other." The amount of peer influence depends on the amount of interaction among the group as well as the nature of their relationships to each other. The amount of contact workers have at work can depend on the design of the work flow and the nature of tasks performed. Some tasks require teamwork while others are accomplished by lone workers. The more interaction, the more the culture influences common practice. How much day-to-day contact do your workers have with fellow workers?

The interaction develops a life cycle over time, and the older it is, the more defined it becomes. As workers interact over years or decades, their impact on each other grows, and the common practices are better known and shared. Even the points of disagreement can become cultural and help to define subcultures or tracks of common practice. If workers with frequent contact perform different tasks, what becomes common or cultural might be more thought than action. This can be especially true of the shared view of safety as a priority or value in the organization.

Certain kinds of work actually foster very close relationships among those who do the work. Underground miners, firefighters, law-enforcement officers, military special forces, and many other groups tend to share common dangers and experiences and view themselves as part of a group (pick your term: brotherhood, profession, fraternity, elite corps, special unit, even "culture") that outsiders do not fully understand or appreciate. Within such groups, the influence of peers is greater than that of outsiders, and peer influence tends to create the shared, unwritten laws by which members reverently abide. We have generalized over our experience that jobs with greater or more specific risks tend to foster these kinds of cultures. How fraternal is your safety culture? Would your workers respect one of their own or an outside expert more?

Contact away from work in the community or other organizations can also impact the culture at work. Cultures in which all workers come from the same community can be significantly different than cultures made up of people from different communities. Workers who contact each other both on and off the job and share both work and home issues can become extremely close. Even if they do not agree on everything, they have common talking points that span across the major portions of their lives. Communication is greatly facilitated when people share common issues. How much contact do your workers have in the community?

Membership in work organizations such as unions or crafts can also impact the ways in which workers interact. These organizations often have their own terminologies, philosophies, and issues that tend to norm the workers and give them talking points. They also have meetings that physically get the workers together more often than just at work. Affiliation is also a norming influence. Many take pride in belonging to these organizations and associate their personal identities with that membership. Others who belong to the same organization are often viewed as closer relations than simply coworkers and culture has a stronger influence on group and individual behaviors. Are your workers connected by union, trade organization, or other affiliations?

Project work such as construction also has a special impact on culture. The workers on a project team often come together suddenly, work together intensely, and then disband. In these environments, cultures tend to either form quickly or not at all. Organizations that do project work usually have learned how to get workers up-to-speed quickly to complete work as contracted. If safety is a priority to such organizations, they have usually also found ways to select workers familiar with their type of work and/or to quickly inform them of the risks and rules needed to keep them safe. Project work can also involve more than one contractor. Communication and correlation of safety efforts among workers from different organizations can also challenge the formation of an effective safety culture. Highly safe project work is a challenge for many organizations, and we will later share some stories of how we have helped to meet this challenge. Is your culture a "project" culture?

The design of work stations and process flow can also either restrict or encourage worker interactions. New employee orientations and teaming, partnering, or mentoring programs also impact how workers contact each other and the roles in which they do so. A new worker may get indoctrinated quickly into the organizational culture or may have to find their way in slowly over time depending on how worker contact is designed. Many organizations lose valuable opportunities to shape culture by failing to design jobs for worker interaction rather than just to get the product out the door. Could a few design changes give you a closer culture?

We learned relatively late the importance of getting workers involved in safety with participative programs. Organizations that utilize some specific involvement-based safety programs such as OSHA's Voluntary Protection Program (VPP) or various versions of BBS sometimes formulate culture around such participation and methodology. VPP can give structure to a culture where various people serve on committees to improve specific aspects of safety, and others can give them input and make suggestions. Some forms of BBS teach a way to approach a fellow worker with a safety concern or how to give positive reinforcement to a fellow worker for good safety performance. Such structure and practices can either become cultural or can cause a reaction within the culture that shapes desired structure or practice in opposition to these models. Either way, the process has impacted the culture. Also, the acceptance or rejection of such programs can impact the relationship between workers and leaders. When leaders adopt popular and well-accepted programs, the bonds and trust between workers and leaders can be strengthened. Unpopular programs can have exactly the opposite effect. (For a more complete description of BBS and VPP, see Appendix A.) Do you have participative programs that allow volunteerism and participation in safety?

The second influence is the supervisor. Again, this influence varies greatly from organization to organization. A few organizations have done away with supervisors altogether and developed self-directed work teams. Such moves obviously impact the culture by removing the supervisor influence and by increasing the peer influence. Most organizations, rather than removing supervisors entirely, have reduced the number of supervisors over the past years. Many sites that had a supervisor for each department on each shift now have one shift supervisor who covers all the departments. Where supervisors still have regular contact with workers, they often have a strong influence on the culture. Supervisors can define common practice and the values with which the group of workers do their jobs. Workers can be a part of the supervisor's culture or form a counterculture that is influenced by how the group disagrees with their supervisor. Such countercultures are common where trust levels are low between workers and supervisors or where supervisors overuse discipline. Do your supervisors work to build a strong culture or do they exercise command-and-control methods?

The third influence is management and/or leaders. Managers and leaders can be the same in some organizations and quite different in others. Managers can lead, and leaders can manage or these two functions can be divided in several ways. The leader's influence is often a distant one removed from the day-to-day reality of the workplace. Even though leaders do not necessarily have regular, direct contact with workers, they may largely control the climate of the organization. The climate impacts the way the culture grows in a broader sense and sets the parameters within which all other cultural influences operate. Many leaders tend to think that they can shape a culture by decree. They can certainly impact it that way, but their decrees are not the only influence on the culture, and one new rule or policy does not necessarily undo all the other influences that shape the culture. In the chapter on climate, we will define the four major areas that leaders should control to allow the culture to grow to its full potential. For now, suffice it to say that management is not the only influence on safety culture and therefore not the only starting place when beginning to shape or reshape the safety culture. Any time the goals of management and the goals of safety are not in alignment, safety will be a subculture and not the overall organizational culture. How visible are your leaders and managers? How well do they communicate their views and exert their influence in the workplace?

The best tool for leadership/management to impact safety culture is the development of a safety strategy. Most organizations have safety wishes or goals; few have true strategies. The development and deployment of a safety strategy is a way for management to impact decisions and practice even from a distance. It can create uniformity of effort and method. It can give a sense of purpose and control to safety efforts and can provide new metrics for helping individuals and groups to better measure their own efforts, contributions, and overall performance in safety. Seeing progress in these metrics can motivate, as well as better direct, improvement efforts. Conversely, the lack of a safety strategy can rob the organization of direction, motivation, and uniformity and rob managers of any real or direct impact on safety efforts. Does your organization have a safety strategy? Is it effective at impacting safety thinking and workplace decisions? Does everyone know the details?

Another complication to accurately defining culture is that most organizations (or even sites) do not have just one. Even if they have one, the one is subdivided into several faceted groups or subcultures. A small manufacturing site may have an overall culture but have a subculture among maintenance workers, production workers, logistics workers, and so on. A service organization can have an overall culture but have subcultures among inside sales, outside sales, support staff, and so on. Organizations can also have completely different cultures not connected by much in common. Multinational organizations can have radically different cultures at sites in different parts of the world due to national or regional cultures. However, a safety culture can potentially have more similarities than overall organizational cultures. Since the issues around safety are more universally similar and less complex, an organization can create a culture of safety that allows for individual or international differences while staying true to some universal values and practices. Do you have safety subcultures? Can they work together or are they (and/or their safety issues) too different?

Cultures have another important characteristic that is often ignored. We call this characteristic the culture's "course." By course, we mean the path through which workers enter into the culture, participate and mature in the culture, and exit the culture. Picture a pipeline with new employees entering in chronological order on one end and mature workers retiring and exiting on the opposite end. In the middle, workers are in various stages of impacting and being impacted by the culture. Many cultural practitioners picture a culture as a box or balloon with members of the culture on the inside and everyone else on the outside. The problem with this model is that it tends to view everyone in the culture in much the same way. The pipeline model acknowledges and illustrates the various levels of cultural indoctrination. These levels more accurately identify stages of culture and prescribe different ways of dealing with people in these stages. An illustration of this model of The Course of Safety Culture™ is in Figure 1. Even cultures that resist change can be changed by starting with new employees and by letting the change flow through the culture

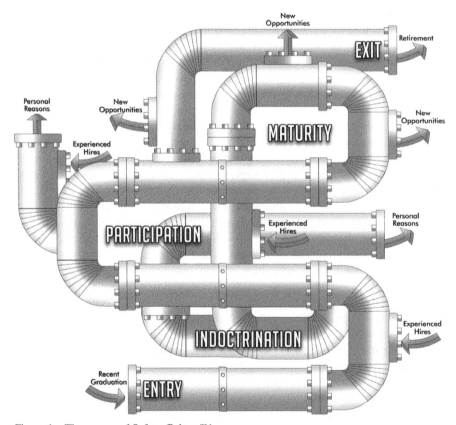

Figure 1   The course of Safety Culture™.

during the course of careers. More resistant workers are usually at or near the end of the pipe and will flow out of the culture over time. We will use this model to suggest potential strategies to deal with cultural challenges in later chapters. What percent of your workers are at the various stages in the cultural pipeline?

But we have only been "describing" culture. Much of the work on culture so far attempts to define the characteristics that the culture should have. Many define an ideal, perfect, or total safety culture and challenge the organization to shape itself to this model. While being *descriptive* has value, being *prescriptive* will more likely empower the organization to move forward. So, instead of simply describing the characteristics of a culture, we will also define the capabilities the culture should develop. Many of the characteristics will be addressed by controlling the climate and chemistry in which we want the culture to grow. Cultures are a dynamic, multifaceted, and constantly changing set of influences that shape common practice. But common practice is the way of doing work; and work can be done better if skills are improved and adopted by the culture. In other words, cultures can learn to improve themselves and develop the common skills and techniques to do so. So, instead of describing what a culture should be, we will develop what a culture should do. And a "can-do" culture can do almost anything, including become excellent at safety. Do you currently have a "can-do" safety culture or a "do-as-you-are-told" culture?

As you consider the answers to these questions, you begin to see how your journey to Safety Culture Excellence will need to be customized. This approach to impacting culture (i.e., infusing the specifically needed skills and methodology vs. defining ideal characteristics and prodding the culture to develop them) has several advantages:

- It prescribes a more direct way to impact the culture (build capabilities).
- It helps define the culture more specifically (what it can do rather than what it is or has or perceives).
- It provides a more discrete metric by which to measure the development of the culture.
- It enables better prediction of how the culture will impact the decisions and common practices of its members.
- It builds the basic skill of improvement into the self-perpetuating fabric of the culture and ensures continuous improvement.

But simply planting the seeds of capability within the culture does not ensure that these seeds will grow. We must also control the climate and the chemistry. If the climate is not conducive to growth, the capabilities of the culture will not reach their full potential or perhaps not even survive. If the basic chemistry of the organization does not provide the necessary elements for growth, the capabilities may die or wither. Safety culture is much more organic than most of the models recognize, and the formation of a safety culture is more akin to growing a plant than to drawing an organizational chart. If you plant the right seeds of capability and control the climate and chemistry, you will grow a safety culture toward excellence. Once it is growing, you can prune and shape it and further adjust the climate and chemistry to maximize its potential. What is your current chemistry and climate and is it conducive to growing excellence into your safety culture?

One last note before proceeding to the basic formula: you are not starting from scratch. You already have a safety culture unless you are building a new facility and staffing it with totally new employees. Any group of people who have been working together for any appreciable length of time have already developed some norms and common practices. Your challenge is not to begin or create a safety culture; it is to improve one. Again, real improvement usually involves improving capabilities rather than transforming characteristics. You must improve the culture's performance, not its nature. Before you try to improve your culture's capabilities, understand what and where it already is and do not reinvent any wheels. Almost all safety cultures have already developed some strengths that can be built upon, and it is always easier to build on existing strengths than to correct weaknesses or flaws. In fact, if you want a positive "can-do" culture, building on their strengths is often the best, fastest, and most direct route to excellence. Find what your culture does well, hone those skills, and aim them at the right targets. What strengths does your current safety culture have? Can you focus on strengthening them rather than correcting their weaknesses?

The following shows the basic formula we propose for improving a safety culture toward excellence. The way you answered the questions in this section should impact how much work it will take to accomplish each of these.

1. *Develop a safety strategy.* The first S in STEPS stands for "strategic." It is impossible to set a strategic target if you do not have a strategy. Most organizations, quite frankly, do not have a safety strategy. They have goals, wishes, programs, and metrics, but all these do not really have a framework that brings them together. An effective strategy can help focus on the right, measurable goals of achieving excellence rather than avoiding failure.

2. *Perform an assessment of your starting place.* Determine what kind of safety culture you already have, what strengths can be utilized, and what additional capabilities it needs to improve. Understand your starting point and use it as a baseline to measure further improvement.

3. *Create clarity of purpose.* Deploy your safety strategy, organize and train the members of the culture at every level in the strategy to learn the basic definitions of safety and the improvements needed. Especially teach the culture the basic skill of targeting and accomplishing what we call STEPS. Share the rationale for improvement, how the organization will benefit and answer the WIIFM question. Structure a Safety Excellence Team (SET) to steer the organization through the STEPS.

4. *Create the right safety climate.* Create or improve the organizational climate in which a safety culture can grow into its personal best.

5. *Create the right safety chemistry.* Make sure that the culture has the elements necessary for safety excellence growth and that these elements are renewed as they are utilized.

6. *Create the control to address the issues of conditions and common practice that impact safety.* Prioritize and address your safety issues one at a time.

7. *Your safety culture can now continuously improve safety.* Reassess, measure, and adjust—recognize progress and barriers and react appropriately and flexibly to meet the changing needs.

Each of these elements represents a milestone that you will reach on your path to Safety Culture Excellence. Each milestone has a series of STEPS that lead to reaching it. As you progress, you may find that you already have some of the STEPS in place and can skip them without compromising your progress. It is ok to do so as long as your efforts meet the criteria of the STEP. Every culture is different, and they each have a unique set of strengths and weaknesses. If you have a strength that helps you skip a STEP you might want to recognize and/or reward that strength and use it to build motivation and momentum for your process. You will find that STEPS focuses on building strengths rather than just correcting weaknesses. This is by design. Safety cultures do not become excellent by simply being less bad. True excellence is achieved by building on existing strengths and by developing new ones.

MAKING THE DECISION TO PURSUE SAFETY CULTURE EXCELLENCE^{SM}

Whenever decisions are made strictly on the basis of bottom-line arithmetic, human beings get crunched along with the numbers.

—Thomas R. Horton

Today's world is full of competing priorities. Leaders of organizations are charged with deciding which priorities should take precedence at any given time. If you are to pursue the course to Safety Culture Excellence^{SM} proposed here, it is essential that the leaders of your organization enter this path with a good knowledge of where it is leading and make a conscious decision to pursue it. It would be ideal for every leader to read this book completely and meet to decide whether or not to proceed. However, in acknowledgement of the lack of ideal circumstances in many organizations, we propose that you can alternatively give a shorter reading assignment or hold a workshop on the materials in this chapter and facilitate the decision.

The decision can be to proceed, not to proceed, or not to proceed right now. If leaders decide not to proceed right now, it is wise to discuss the potential time to start and to set a follow-up date on which to discuss the matter again. If your leaders decide not to proceed, there are other ways than a top-down approach to address Safety Culture Excellence. Unfortunately, a thorough discussion of those options is beyond the scope of this book. If your leaders decide to proceed, the path is spelled out in the rest of the book, and you can proceed at your own speed through the remaining STEPS.

If you decide to assign the reading of this chapter and have a meeting to decide, ask leaders to jot down their answers to the questions in each exercise and bring them to the meeting. Use the flow of these exercises to direct your discussion in that meeting. If you decide to hold a workshop, you should turn these exercises into brainstorming sessions and plan to capture the answers to the questions in some way that they can remain visible and/or accessible to the leaders as they make their decision. Even if you are planning a workshop, you can assign the reading of the chapter and taking notes as preparatory work, which usually facilitates the flow and speed of the workshop.

However you plan to proceed, make sure that the meeting or workshop stays focused on the potential for positive gains and does not become a fault-finding critique of current or past safety efforts. The STEPS process is designed to be a logical next phase of safety progression and not a remedial solution to failed efforts. The focus of this decision-making session should be the path forward rather than history. Encourage everyone to give up all hopes of improving the past and look to the future.

The organization of this chapter is basically a series of exercises in which participants or readers are asked to review background information and then consider questions about the topic. The questions are designed to aid both analysis of existing efforts and strategies and to suggest how those could potentially be improved. There are no right or wrong answers to the questions. This is not a quiz but a thought-provoking exercise. The goal is to arm leaders to make an informed decision of whether or not to move on through the other STEPS in pursuit of Safety Culture Excellence.

The flow of these exercises and questions is as follows:

Exercise 1: Understanding and Breaking Out of the Cycle of Avoiding Failure

Exercise 2: Safety Culture and Performance Excellence Strategy

Exercise 3: Prestrategy Reality

Exercise 4: Starting Point Evaluations

Exercise 5: Forming and Norming Culture

Exercise 6: The Bridge to Safety Excellence™

Exercise 7: Making the Decision to Move Forward

EXERCISE 1: UNDERSTANDING AND BREAKING OUT OF THE CYCLE OF AVOIDING FAILURE

Background Materials

Most organizations are trying to avoid failure in their safety efforts. They know very well what failure looks like. It is painful and expensive. It hurts the employee, the morale, the public image of the organization, productivity, and profits. It seems like a very good thing to avoid. Along with this thinking is the assumption that success is the lack of failure. If you avoid all the negatives, that is positive.

There is a parallel to this situation in health care. If a patient thinks he or she is healthy simply because they have no recognizable major illness and no severe pain, they might not seek medical help. Regular checkups are often an effective way to detect health problems before they become deadly or untreatable. The fact that symptoms are not obvious does not mean they do not exist. Risks, like disease symptoms, can also appear dormant until it is too late. Lack of disease is not necessarily excellent health just as lack of accidents is not necessarily excellent safety. Thinking of health or safety as a vacuum in which undesired things do not happen is both unrealistic and unattainable. It also fosters a "wait-and-see" mentality in which we assume that everything is ok until the undesired event happens; then we react. In medicine, a reaction is an undesirable event. A patient who reacts to a medicine or transplant is in danger. The desired event is for the patient to respond, not react. In safety, we should also respond to our safety strategy rather than react to accidents.

This thinking is reinforced by the way most organizations measure safety. The pervasive use of lagging indicators tends to drive a "golf score" mentality in which the fewer the strokes the better the score. Although rates (recordable rates, severity rates, etc.) serve the purpose of providing an equitable comparative metric between organizations of different sizes, they also constitute a measurement of failure. When you manage with failure metrics, you tend to set goals and develop plans to "fail less." Failing less is not a strategy and also not a good definition of success. It is certainly a goal and a desire for any caring organization. However, wanting to fail less does not prescribe how to do so. Failure metrics can provide lessons learned to avoid the repetition of failure, but these are still short of a definition of true success. Many organizations tell us that they are trying to move from lagging to leading indicators in safety. Perhaps a better way of thinking is to move from failure metrics to success metrics.

This cycle of avoiding failure is counterproductive to developing a culture of safety excellence. Some of the characteristics of the "avoiding failure" mentality include the following:

- It tends to drive reactive rather than proactive efforts.
- It clouds the vision of what true safety excellence looks like.
- It leads to short-term thinking and accountability.
- It drives a "program" mentality of adopting the newest fad to fail less.
- It encourages numeric improvement goals that, in turn, may encourage pencil whipping.
- It discourages process indicators in favor of impacting the all-important lagging indicators.
- It encourages "creative interpretation" of the lagging indicator data to justify performance.
- It measures what you do not want but fails to measure what you do want.

The cycle that is created by this type of thinking is described in Figure 2. We call it the "perpetual cycle of avoiding failures." It begins with reviewing the incident rate data and setting goals for reducing the incident rates to a new level. This targeted lowering of the failure rate drives the development of initiatives to accomplish the goal. Often, the relationship between these initiatives and the goal are unclear, uncertain, or even missing. But, a goal necessitates effort to reach it, and programs and other efforts that have safety in the title can create the perception of being aimed at the goal. Even if the initiatives are well designed to achieve the desired improvement, they are not always carried out effectively and often lack process metrics to measure the effectiveness of the effort. The plan is for the initiatives to impact the incident rate. This is how they ultimately prove their effectiveness and help the organization realize their goals.

The problems with beginning and ending the process with lagging indicators are largely responsible for the failure of many such plans. There are so many factors

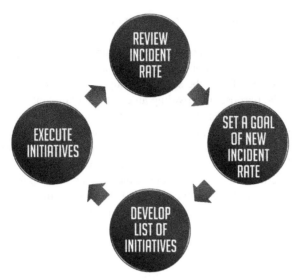

Figure 2   The Perpetual Cycle of Avoiding Failure™.

that impact incident rates that it is hard to determine if one new initiative is solely or partly responsible for the changes. Normal variation of the rate is often used to justify initiatives or to reward efforts when they have little or no statistical significance. Initiatives are often abandoned when they are just beginning to work because they take so long to impact the lagging indicator. Also, since many organizations look mainly at the incident rates (or recordable rates), they may fail to see if the initiative is impacting severity rates, costs, or other lagging indicators. We have seen initiatives that were beginning to make significant differences in the safety culture abandoned because they had not yet significantly changed the lagging indicators.

The way to begin to get out of the mentality of avoiding failure is to begin to climb the "Safety Performance Excellence Curve™." This journey has usually already begun with safety edicts such as legislative requirements that become rules and procedures. These edicts set a basic level of expectations for safety performance that, if reinforced by management and supervisory enforcement, get the workers "hands and feet" performing at a safer level. We use the term "hands and feet" to indicate that these actions are usually based on requirements and do not involve workers at the motivational level. They do them because they are required, not because they believe in them, buy into them, or understand how they are important. They either result in the required minimum effort from the worker or in avoidance behaviors. These are efforts to not get caught in noncompliance. (For instance, do employees always wear their PPE or do they only do so when the supervisor is on the shop floor?)

To move the organization from this basic level of safety performance to something more perfect, it is necessary to set goals for improvement, not just compliance. It is also critical that employees know the rationale for such improvement. How will such improvement provide additional value to the organization and to the individuals in it? How will this approach to improvement be a good fit for the organization and help to make improvement easy and natural? For some organizations, safety improvement will help to create competitive advantage. For others, it might help to land contracts with better clients. Still others might need to improve safety to improve public relations or to be able to recruit better new employees. Whatever the rationale, it needs to be shared with the employees to get their "hearts and minds" into the effort along with their "hands and feet."

The late Stephen Covey in his famous book *Seven Habits of Highly Effective People* said, "Hands and backs can be bought, hearts must be won." Workers will do the basics to get a paycheck. If you want them to go above and beyond the basics, you need to win their hearts over to the quest for safety excellence. Hearts are not won by understanding what is to be done; they are won by understanding why it is important and how it will impact themselves, their fellow workers, and the organization. Workers with their hearts in the effort perform at a completely different level than those who are simply compliant. Just like a sports team with great talent can be beaten by a team with their hearts in the game, an organization with heart can outperform an organization with compliance as their only goal. Excellence begins with the reason for excellence. Once the "why" wins the hearts, the minds will figure out the "how." This progression is illustrated in Figure 3.

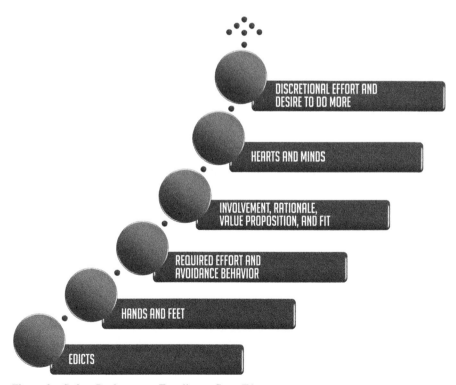

Figure 3   Safety Performance Excellence Curve™.

Questions for Exercise 1

1. To what extent do we rely on lagging indicators to measure and manage safety?
2. Do we set goals based on impacting the lagging indicators?
3. Could these goals encourage or reinforce cheating?
4. Do we measure the success of safety initiatives strictly by how they impact lagging indicators?
5. Do we manage safety with edicts and aim mainly at compliance from workers?
6. Are we getting the minimum performance from workers in safety or do they take initiative?
7. Do we communicate the rationale of safety to our employees effectively and convincingly?
8. Do we have meaningful and rewarding involvement opportunities for our workers in safety?
9. To what extent is the average worker creatively and emotionally involved in our safety efforts?
10. How much more effort could and would our workers put into safety if we effectively solicited their discretionary efforts?

EXERCISE 2: SAFETY CULTURE AND PERFORMANCE EXCELLENCE STRATEGY

Background Materials

The development of a safety strategy by the organization's leaders involves 11 different aspects. We call these, together, The Path to Safety and Performance Excellence Strategy™. The leaders should learn and start to become familiar with these elements in this first workshop and then develop them more fully in the second workshop. The 11 aspects will be introduced, and the questions for this section will focus on the prestrategy reality of the organization. For an illustration of the flow of the 11 aspects, see Figure 4. Each element of the strategy is described as follows:

1. *Purpose.* The purpose for safety excellence is the rationale we discussed in the previous section. Why exactly does the organization believe that safety excellence should be pursued? What is the business purpose of such an effort and what are the anticipated benefits if successful? Why does this effort belong with other organizational goals and how do they work together for the overall good? It is not only important to intelligently answer such questions but also to share the answers with all the employees whose efforts will be required to be successful. A statement of purpose should be motivational and appeal to the hearts and minds of the recipients. It is not just the reason for the decision; it is the greater good that will result.

2. *Core values (different than situational values).* What does the organization truly value? How important is the well-being of employees and associates

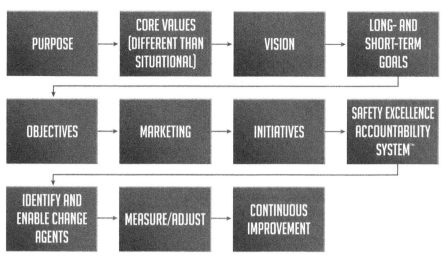

Figure 4   Safety Culture and Performance Excellence Strategy™.

compared with other values such as profits, shareholder values, accomplishment of mission, and so on. Statements such as "begin every job with safety" usually represent situational values. Situational values tell employees how to make decisions in the workplace under specific circumstances. Core values represent deeper levels of organizational purpose. How does safety rate against these other core values in every situation? Also, what are the core values of safety? What beliefs should members of the organization share around safety and its importance? It is hard to make beliefs shared within the culture if they are not stated and reinforced.

   Case Study: One of the authors was working with an organization to create a clear, motivational, and measurable vision statement to direct the strategy. While preparing for the workshop and reviewing the core values of the organization, we discovered that safety was not among them. It is hard to create a vision of safety excellence if safety is not inferred or explicitly stated as a core value. Sometimes the opportunities for improvement are surprisingly obvious when you are looking through the right improvement lens.

3. *Vision.* What does safety success look like? If you could not see the OSHA log, what else would you look for to tell if the organization was excellent in safety? What would people know? What would they focus on? What metrics would indicate success? The answer to these and other important questions can help guide leaders to form a vision of safety excellence. It is critical that this description not just be results (lagging indicators) but the things that produce the results. What processes would be in place and how would you know they were working, other than from a lack of failure (accidents)? Even if accidents have greatly reduced or gone away, what would you see people saying and doing that lets you know how and why you have great results. To determine this, you need to think what safety excellence should look like 5 years from now. Do people see and comprehend how their actions produce the results? Do they feel their contributions are recognized and appreciated? Even if we achieve excellent results next year, if we do not know what it looks like and how it works, will we be able to reproduce it the year after next? If not, we are managing by voodoo and simply hoping that our good fortune will continue. Much of the answer to these questions may be contained in the items that follow.

4. *Long- and short-term goals.* Almost all organizations set goals for results, but not always for the efforts and initiatives that are going to produce the goals. All goals should be aligned with and play an integral part in the overall strategy. Goals often need to be stated in a prescribed order to make sure that they build on each other rather than conflicting or attempting to build on structure not yet established. Short-term goals should not dominate and should be aimed at attaining longer term goals. There are quick fixes and ideal solutions, but there are also a lot of goals that fall between those two extremes. Goals should have targeted completion dates, but everyone should remember that we cannot always accurately predict the future. While aiming at goals, it is important to be flexible without being lax about completion dates. Goals need to provide measurements of progress. Goals are not either achieved or not achieved; they are some percent achieved. Progress is the most important aspect of excellence. Everyone should see visible progress toward goals often if they are to be continuously motivated toward excellence.

5. *Objectives.* These are the reasons behind the goals; the states to be reached and the benefits of reaching them. To an extent, goals are what you achieve and objectives are what you become from reaching them. To reach objectives, you truly must begin with the end in mind. Objectives can also be defined as parts of your vision. When you break down what success looks like, that might be a series of objectives. Additionally, the objectives provide a clear outline of what needs to be accomplished, why and how it will be measured, providing a sense of strategy to the lower levels in the organization responsible for the tactics.

6. *Marketing.* This is a foreign term to many involved in safety, and we are often asked what this concept has to do with safety. To fully understand the answer to this, you must think of workers as the customers of safety programs and efforts, not the problems. When you think this way, it becomes clear that the programs and efforts must be marketed to the intended customers. Workers must "buy in" to the safety efforts if we want more from them than compliance. We have asked safety professionals if they have ever washed a rental car and why not. The answer is usually obvious:" the car doesn't belong to me!" Neither do safety programs that have not been marketed and "bought into." Workers might use them, but they will not feel a sense of ownership or pride, and they will not take the same care of them as they do of things they own.

7. *Initiatives.* Nothing is accomplished without effort. What specific efforts will be required to accomplish your safety-improvement goals? What existing programs might help and what other initiatives need to be developed to get your organization from where it is to where it wants to be? These are the initiatives you need to fulfill your safety strategy and help you achieve safety performance excellence.

8. *Safety Excellence Accountability System™.* Everyone in the organization needs to understand his or her roles, responsibilities, and results they must accomplish to help the organization succeed. Each person should also be accountable for fulfilling these roles, responsibilities, and results (RRRs), and the system of accountability should be established, understood, and followed. There should be consequences for fulfilling these and for not doing so. These should be known by all, and there should be no surprises when accountability is exercised. This is an opportunity to set and reinforce clear expectations for individual contributions to overall organizational safety excellence.

    (a) Roles are what individuals should "be." Leaders might have the role of "resource provider" or even "cheerleader" of safety. Supervisors might be asked to be "guardians" or "overseers" of individuals. Workers might be assigned to be "examples" to their fellow workers and "mentor" to contractors. If you are fulfilling your roles, the title of that role should be the way you are perceived by others.

    (b) Responsibilities are what individuals should "do" to fulfill their roles. It might take multiple responsibilities to fulfill a safety role. Leaders might be asked to "communicate safety values with his or her direct reports." Supervisors might be asked to "discuss job-specific safety issues in all preshift meetings." Workers might be asked to "know or look up the PPE requirements for each new job before you begin work."

    (c) Results are what individuals should "accomplish" if they are fulfilling their roles and responsibilities. Leaders might be expected to "enable everyone in the organization to articulate the value of safety." Supervisors might be expected to "maintain a sense of vulnerability among workers and set a focus and feedback system for safety improvements." Workers might be expected to "create teamwork in safety where it is accepted and expected that workers warn each other about potential job hazards and share precautions to minimize dangers."

9. *Identify and enable change agents.* To accomplish successful improvement, someone needs to be in charge. We encourage the change agents to be more than one, that is, a team or committee (whichever terminology fits your organization best). In subsequent chapters, we are going to suggest that you form a SET made up of a cross section of employees to spearhead and steer the change efforts. Not everyone can lead, so it is important that the change agents be viewed as representatives of the entire workforce and that every employee feel represented on this team. The next chore is to enable these people to really make the change happen by providing them with the training, time, resources, and support to make the desired goals a reality.

10. *Measure/adjust.* Even the best of plans can change when they meet up with reality. The famed boxer Mike Tyson once quipped, "Everyone's got a plan until they get punched in the face." The U.S. Military advises, "All strategies are successful until contact with the enemy, because the enemy hasn't voted yet." The workplace has certain characteristics that impact any and all change efforts. These characteristics are difficult to identify and even more difficult to predict how they will impact change efforts. When you take a plan to the real world, some aspects tend to work and others do not. It is critically important that you measure the success or failure of your plans and adjust quickly to any obstacles or barriers that get in the way of success. To do this, it is important to develop measurements of critical aspects of your plan that will allow you to monitor their progress. Measuring human efforts is a challenge, and it is important to measure what is important, not just what is easily measured.

11. *Continuous improvement.* Good plans move from one improvement to another. There is no destination except continuous improvement. Excellence, as we have already stated, is a journey and not a destination. Organizations are either improving or sliding back. There is no way to really maintain the status quo in safety. Workers are either more or less aware of risks. They are either taking adequate precautions more often or less often. It is important not to develop a mindset of "maintenance" or of "good enough." In safety, there is no real stasis. Efforts and results are always moving. The challenge is to keep them moving in the right direction.

EXERCISE 3: PRESTRATEGY REALITY

Background Materials

Every journey has a starting point. Your safety efforts have a current status, and determining that status is important to do briefly during this workshop and more thoroughly as you actually begin to apply that strategy to the STEPS that will lead your culture to excellent safety performance. For right now, look at the elements on Figure 5 and follow the descriptions of each element that follows.

We often ask participants in our workshops to describe, in their own words, the current status of their safety efforts. It is interesting to see not only how the descriptions differ but also how one impacts the others. Synergy is the impact of people on each other, but so is "groupthink." It is important to synergize without intimidating. When one person describes the status, you can see others modifying their descriptions to include issues they had not before considered or they modify their evaluations based on the evaluation of others. Almost all evaluations can be more complete and uniform if they are focused on the same aspects of the subject. For this evaluation, we have chosen to focus the evaluation of the current situation

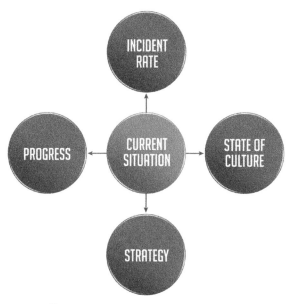

Figure 5   Prestrategy reality.

on four aspects: incident rates, state of culture, existing strategy, and the view of progress.

Questions for Exercise 3

Incident Rates

1. What is your current incident rate (total recordable incidence rate [TRIR]) and severity rate?
2. How does that compare to your industry code average?
3. How are your rates trending: improving, staying level, or getting worse?
4. How have your rates responded to other efforts to improve them?
5. How many of your employees could accurately quote what your rates are?
6. Do they understand how their day-to-day actions impact the rates?
7. What special causes have impacted your rates: layoffs, new hires, changes in the workplace, and so on?
8. What improvement goals have you set for incident rates, what initiatives have you used to achieve them, and what was your success rate?
9. How does your current safety status impact your ability to do business, to land good contracts, to recruit great talent, and so on?
10. What is your rationale for dedicating effort and resources to further reduce accident rates?

State of Culture

1. What is your existing safety culture like? How do you know?
2. Have you measured perceptions or held interview sessions with focus groups?
3. What is common practice and how does it change if managers and supervisors are not present?
4. How do your leaders lead safety and how do your supervisors supervise?
5. Is your management style "command and control" or is it more empowering?
6. Do your workers understand how safety is accomplished and is their role clearly defined and reinforced? How do workers interact with each other and what is their tolerance for risks?
7. At what point would a worker intervene with a fellow worker to prevent potential injuries?
8. Do your supervisors know how to help their workers improve in safety? Are they safety cops or coaches?

Current Strategy

1. Do you have a safety strategy?
2. Who helped in the creation of the strategy?
3. Have you validated knowledge transfer of the strategy to all levels of the organization?
4. Is there a sense of ownership and belief in the strategy at all levels and all areas?
5. Is the strategy behaviorally actionable by any level in the organization?
6. How many times, by level, is the strategy behaviorally reinforced throughout the day?
7. Is there a balance of consequences (positive and negative) for those who contribute and those who act in opposition of the strategy?
8. What motivators or demotivators exist in your systems and culture that might positively or negatively influence the strategy?
9. If someone were to quiz your workforce to recite from memory the safety strategy, how many would be able?
10. Do you have improvement goals?
11. What is your plan to accomplish your improvement goals? Is it aligned with the goals or simply hopeful? Are your leaders thinking about safety strategically or tactically?
12. Do you have the program-of-the-month mentality in safety?
13. What is your definition of safety? What is your definition of an accident?
14. What percent of your workers share your definitions and can quote them?
15. Does everyone know their role in your safety strategy?

16. Do you have their hearts and minds or just their hands and feet? What do you think your probability of success is with your current strategy?

17. Do you wish you had a chance to start over?

Progress

1. How is your organization doing in its safety performance? Are you improving, maintaining, or backsliding?

2. Do you have a formula for progress?

3. Are your leaders bent on improvement or simply "going with the flow?"

4. Do you have a history of improvement or do your results ebb and flow?

5. Do you feel in control of your results or do they seem to have a life of their own?

6. Is progress your constant goal or would you be happy to maintain your current status?

7. Do you feel you have a handle on how to improve or are you the victim of changing circumstances?

8. Do you view safety as a systems issue or do you blame individuals for the failures in your organization?

EXERCISE 4: STARTING POINT EVALUATIONS

Background Materials

As we look at these four points of your pre-strategy reality, let us also look at the three levels in your leadership ranks and evaluate their current status in regard to improving safety. In the workshop to develop your safety strategy, it is important to know your starting point. This exercise is not aimed at blaming or fault finding, it is aimed at evaluating and developing effective improvement strategies. So, evaluate your executives, managers, and supervisors one at a time and determine based on your current perceptions if each of them is inspiring and driving positive change of simply maintaining the status quo. Turn you individual evaluations into percentages of the whole number in each category and fill in your percentages using Figure 6 to determine your starting point for your new safety strategy.

Questions for Exercise 4

1. What percent of your executives inspire and drive change?

2. Who do they report up through?

3. What percent of your executives maintain the status quo?

4. What percent of your managers inspire and drive change?

5. Who do they report up through?

6. What percent of your managers maintain the status quo?

Figure 6   Starting point evaluations.

7. What percent of your supervisors inspire and drive change?

8. Who do they report up through?

9. What percent of your supervisors maintain the status quo?

After you develop your safety strategy, we are going to suggest that you do a thorough assessment of the starting place of your safety culture. This will give you a baseline against which you can measure your progress in Safety Culture Excellence as well as help determine your best path forward. This evaluation is different from determining your starting place for strategy development in which you look at limited elements. The strategy will help you to move forward quickly with purpose, and the cultural analysis will help you to determine the strengths you can build on and the issues that might challenge your progress toward Safety Culture Excellence. The assessment will also serve as a baseline to measure your progress.

EXERCISE 5: FORMING AND NORMING CULTURE

Background Materials

To better understand how to improve your culture, it is important to understand how the culture reached its current state. The following diagram, Figure 7, shows the flow of cultural elements that impact the shared values and perceptions that form within the group. When people work together, they begin to share experiences and information and associate with each other and the organization's leaders, managers, and supervisors. These factors begin to form an impression of the organization that impacts both feelings about the organization and a sense of "how we do things around here" that forms the cultural norms. The experiences everyday either reinforce or challenge these perceptions. As the perceptions are reinforced or modified, they form the impressions about what the culture is like and how to best function within it. Think about how each of these elements works and impacts others as you follow the discussion of this diagram.

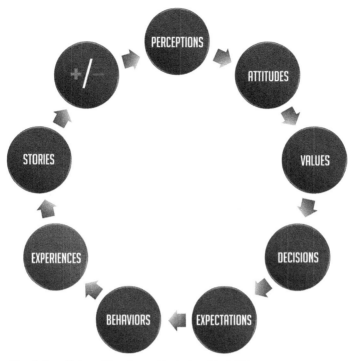

Figure 7   The Safety Culture Excellence Evolution Model™.

1. *Perceptions.* Perceptions are basically the brain's attempt to summarize the information coming to it from the senses. One person's perceptions can vary from those of another based on different inputs but also based on different interpretations of the same input. People who work at the same place around the same people will generally develop some common perceptions. Where these perceptions overlap, we begin to see culture.

   One of the complications of working with perceptions is that they fall into two broad categories: accurate and inaccurate. We have heard many people say that "perception is reality." That may possibly be true of accurate perceptions but is definitely not true of inaccurate ones. Although perceptions tend to be viewed as reality, the simple fact is that human beings can perceive things to be what they really are not. Inaccurate information can cause inaccurate perceptions. Misinterpretation of accurate information can cause inaccurate perceptions. Even personal beliefs and prejudices as well as emotional responses can cause inaccurate perceptions. The real importance of determining whether or not perceptions are accurate is because you change inaccurate perceptions a different way than you change accurate ones. Inaccurate perceptions can often be changed simply by sharing more accurate information and by clearing up misunderstandings. Changing accurate perceptions often involves changing the reality of the situation being perceived. For example, if your workers perceive that your safety training is not useful in helping them do their jobs safely, do they need more information about the design of the training or do you need to redesign the training to better meet their needs?

The way people perceive their reality impacts their attitudes. In almost all organizations, there is a shared perception of what safety is, how it works, and how important it is. The more common these perceptions are, the stronger the culture has become.

2. *Attitudes.* Just as perceptions are the way you think about your reality, attitude is the way you feel about it. Feelings can change or become stronger based on ongoing inputs from the environment and the people in it. The way you think and feel about things impacts what you consider to be important.

People with good attitudes tend to do better quality and quantity of work and enjoy what they do more. Excellence in organizational performance is therefore at least somewhat dependent on attitudes. The problem with trying to change attitudes is that they tend to be a by-product of other factors and are not easily changeable by a direct-frontal assault. People's attitudes usually change as a result of changing perceptions or beliefs rather than from direct efforts to change attitudes. If your workers have a bad attitude about your safety program, you can probably change their attitudes more effectively by improving the program or by sharing better information about the program than by talking to the workers about their attitudes.

For these reasons, we recommend that you consider attitudes an indicator of culture rather than a control point. If you try to control perceptions and create experiences that produce stories that travel through the culture, you will probably be effective in changing attitudes. The noticeable change in attitudes will be an indicator that your efforts to change perceptions and experiences are having the desired impact on the culture.

3. *Values.* People perceive and feel that some things are more important than others. Important things influence us more than things we view as less important. We develop a sense of values both from our own thoughts and from our environment and from others who influence us. Organizations have values as well, both officially and unofficially. What takes priority is perceived to be more important than the things that do not get the same attention. However, there is an important difference between priorities and values. Priorities can and do change whereas values tend to be constants. Values tend to be more universal than priorities, so constantly changing priorities do not often become fixed values. However, over time, both individual and organizational values can be influenced by reinforcing and managing perceptions.

There has been a lot of discussion in recent years about safety and the fact that it should be an organizational value rather than a changing priority. Some argue that the changing of the priority of safety when lagging indicators change actually drives the variability of safety results. For example, when incident rates increase to a certain level, the organization increases its focus on safety and prioritizes safety efforts. Usually, such attention drives incident rates back down over time. If the organization reacts to the lower rates by reducing the priority of safety efforts, that lowering of priority often results in the incident rate climbing again.

The idea is that safety efforts should be maintained even when accident rates do not demand attention. This is directly parallel to the assertion of the "zero defects" or "Six Sigma" mentality in quality efforts. It means that the goal should not include a tolerance for defects. In safety, if the organization has a tolerance for defects (in this case risks and resulting accidents), it will tend to react when the results go beyond the tolerance level and relax efforts when the results are once again within the tolerance range. Excellence is a constant striving for better and better, not an effort to reach an acceptable level of failure. This would suggest that safety must be an organizational and personal value to truly achieve excellence. However, the status of a "value" must be more than a change in terminology. When you ask workers what the importance of safety is and the answer is constantly high, that is a better indication of safety becoming a value. When other priorities challenge safety and yet safety efforts continue, that is the ultimate indicator that safety is becoming a value rather than a changeable priority.

4. *Decisions.* Decisions are influenced not only by situations but also by perceptions, attitudes, and values. They have both personal and cultural influences. The decision to take a risk or a precaution, to report a near miss or not will depend on both how the individual thinks and feels and how their fellow workers and leaders think, feel, and react. Many organizations make the mistake of thinking that all decisions are within the power of the individual and strictly a matter of personal choice. Most decisions are heavily influenced by outside factors, and some are actually beyond the control of the individual. Understanding what influences decisions is key to making better decisions within the culture.

Many organizations make the mistake of judging and evaluating decisions rather than trying to understand how and why they were made and what influenced them. We constantly see organizations that praise workers for production levels without seeing the safety shortcuts that led to the increased productivity. Likewise, workers who take precautions and do work safely are often told to be more like the more productive workers without realizing that that means becoming less safe.

Even when productivity is not the main issue, many organizations do not discover what influences workers to take risks or limits their ability to take precautions until these factors cause accidents. Talking about common decisions in safety meetings and training sessions can impact how these decisions are made. It is also an opportunity to discover factors that influence safety decisions before they result in injuries. Creating experiences that result in stories of good safety decisions can also influence future decisions.

Warning: some organizations have made efforts to eliminate worker decisions from the workplace through creation of procedures, permits, or oversight policies. The control of decision rights in safety is worthwhile, and decisions should be made by the best qualified people whenever possible. However, trying to completely remove decisions from the workplace has proven disastrous on many occasions. Some created an environment where everything stops until a decision maker can be brought on the scene. Others have encouraged workers to rely on procedures rather than to analyze the risks, and this has created a false sense of security that actually caused injuries. As long as we rely on human beings as a part of our processes, we need to equip them with a good knowledge of hazards and the ability to make safe decisions. Trying to completely eliminate worker decisions is seldom, if ever the best approach.

5. *Expectations.* Most decisions are made with the expectation of how the decision will produce results. Expectations can be based on experience, common sense, or the input of others. The organization usually has expectations of its employees, and these can also influence decisions and how the employee expects the organization to react to their decisions.

   "All disappointment is based on a set level of expectations." It is important to remember this old adage. When an employee decides to intervene for a safety concern, report a near miss, or volunteer for a safety initiative, they predict the probability of what will occur following this behavior. Decisions are made based on the anticipated consequences that will follow the behavior in mind. Unless the individual is a glutton for punishment, normal, rational humans do not make decisions knowing they will result in an undesirable consequence. If a worker believes the decisions to help a new employee, conduct a job observation or suggest an innovative new solution will be supported and recognized by the supervisor, the desirable behaviors are likely. Workers who have decided to get involved and had negative experiences as a result are less likely to get involved in the future, unless they hear stories which suggest that things have changed.

   The ability to accurately predict outcomes, that is, how certain decisions will produce results is a key to safety. One definition of an "accident" is making a decision about how to get the job done and having the resulting behavior(s) produce an injury instead of a completed job. Unwanted and unpredicted outcomes are accidents. Many workers who do not get physically injured by their decisions get mentally harmed when the boss or fellow workers do not react to their decisions as they expected. A worker who is asked to report near misses, does so and is punished for his actions just had an accident. It is not the kind that does bodily harm but another kind of injury. These types of accidents need to be prevented also, and this kind of injury will keep a culture from reaching its ultimate goal of excellence if not addressed.

6. *Behaviors.* Behaviors are the visible acts of the people in the culture. They are the results of the decisions made that were influenced by the other factors mentioned. Behaviors produce results called consequences. People anticipate these consequences and try to fashion their behaviors to produce desired ones.

   The term "behavior" has been burdened with a lot of negative connotation. Even outside the professional world, behavior has become negative. A teacher tells a student to "behave" because he or she is "misbehaving." A disease is labeled as caused by controllable behavior. Even when my wife says she wants to talk to me about my behavior at the party last night, I do not anticipate being praised for what I did right. We tend to assume that behavior is usually bad behavior.

The scientific definition of behavior is simply "an observable act." Anything a person does that can be seen is technically a behavior. But the word tends to connote blame rather than action. When we say that an accident was caused by behavior or could have been prevented by behavior, we tend to jump to the conclusion that the worker doing or not doing this behavior was somehow to blame. This narrow view ignores the fact that not all behavior is within the control of the worker and that we do not always accurately predict how our behaviors might produce consequences. It also ignores the many factors that influence behaviors other than individual choices.

In our professional assessments of safety cultures, we seldom find workers deliberately taking risks or knowingly violating procedures unless there are other factors influencing those behaviors. Sometimes the influence is simply a common practice that has developed and become "the way we do this." More often, the decisions are influenced by the location and availability of tools and equipment or workplace designs that are counterintuitive. Many behaviors are influenced by workflow, procedures, or time guidelines that promote rushing and shortcutting. Simply asking workers to be safe while ignoring these factors which make that difficult or impossible is not the path to excellence. The beginning of making behaviors safer is understanding what influences them. If we do not identify and address the influencers, we might not effectively change the behavior.

7. *Experiences.* The consequences of behaviors produce experiences. How others react to the decisions contributes to our perceptions of these experiences. We tend to fashion our decisions to create good experiences, and we measure our level of success by how we perceive these experiences.

Experiences are also a potential control point for safety culture. We can create positive experiences that can potentially change perceptions of those having the experiences and create stories that will impact the perceptions of even more members of the culture. When managers and supervisors set good examples for safety, when they constantly discuss safety, and when they ask workers for suggestions to improve safety, they create experiences. When they stay in their offices and do not make contact, they create rumors.

Positive safety experiences can be created in meetings and training sessions as well. Many organizations have taken the boring meetings and required yearly training to whole new levels by making it an experience in which workers can discuss and input ideas, leaders can share vision and strategies, and cultures can find focus. Excellent organizations manage opportunities to create positive experiences.

8. *Stories.* Not everyone in the culture experiences everything that happens, but significant events are often turned into stories that are told throughout the organization. These stories expand personal experience and can change the perceptions of what is and is not culturally acceptable.

The best way to create positive stories is to create positive experiences; however, it does little good to aid in the creation of stories if they have no way to be communicated. Giving workers opportunities to interact during meetings, break times, and in training sessions can pay big dividends if what they say to each other improves the safety culture. Organizations that stifle interpersonal communication often do so because they know they have created negative experiences and stories and would prefer workers to keep them to themselves.

It is going to be very difficult to reach Safety Culture Excellence in a noncommunicative environment. The more you encourage and empower communication, the more teamwork and focus you can develop. Official channels of communication are desirable for such purposes. Usually, the grapevine (unofficial channels of communication) is inversely proportional to these official channels. Every organization will inevitably have both, but the more you can flow information officially, the more you can monitor it and understand what is happening in the culture and what influences it.

Another important use of stories involves the theory of receipt of communication. When a worker is told to take a precaution, the communication is received in a totally different way than when the worker discovers the wisdom of a precaution. We are finding that stories facilitate discovery learning. We have advised many organizations to communicate accident data through the story format rather than making generalizations or classifications. Simply telling the story of what happened leading up to the injury helps workers better put themselves hypothetically into that situation and second-guess the decisions. Communicating accident data this way almost creates a simulator in which workers can safely duplicate the experience rather than simply receiving information.

Stories, in short, involve workers rather than simply making them the receivers of communication. When the managers and supervisors communicate through stories, they encourage workers to do likewise. When storytelling becomes cultural, communication becomes more engaging. Active versus passive receiving of the message makes it a call to action rather than simply a receiving of information. Stories are also remembered better than facts, and so the information and lessons learned have a longer shelf life. The use of parables and fables was very prevalent in classical literature and early religion. It has a history of effectiveness as a form of communication and can make safety communication more effective as well.

9. +/–. These stories can either reinforce perceptions or change them. These perceptions either become stronger or start over based on the experiences and stories that make up the culture. Very few leaders realize the power of the stories spreading within the culture to shape perceptions. Therefore, few leaders try to manage the perceptions by helping to create experiences that become stories that shape perceptions. It seems an indirect route but has proven to be one of the most powerful and effective methods to shape positive perceptions and to improve performance.

When someone takes action and behaves in a certain way resulting in a positive or negative experience, stories are told to others throughout the organization that either confirm (+) or conflict (–) with the existing individual or shared perceptions. Moreover, negative experiences are known to be spread more virally than positive experiences. The worse the experience, the more people will know about it. Stories are the tribal characteristics of an organizational culture. Whether formal or informal, they are the most effective influence on decisions and behaviors.

Let us take an example and run it through these stages:

- *Perceptions.* Safety is very important to the supervisors here.
- *Attitudes.* I like working where my safety is important. I like it here.
- *Values.* I should always consider safety when I make workplace decisions.
- *Decisions.* I could rush this job, but I think it would compromise safety. I decide to do the job more slowly but safely.
- *Expectations.* I expect that my boss and other supervisors here will support my decision.
- *Behaviors.* I do the job slowly and safely.
- *Experiences.* My supervisor sees me and criticizes me for taking so long to complete the job. He says he wants me to be safe, but we have to get the work done on time.
- *Stories.* I tell my fellow workers what happened, and they repeat the story.
- *+/–.* Everyone who hears the story has an altered perception about the importance of safety among the supervisors. Does my attitude change? Do I change my mind about company values? Do I still feel proud to work here? Next time, will workers who heard the story tend to make a different decision? How does this story impact the company safety culture? What if a lot of such stories are told? What if this is just one instance and all the other stories are more positive?

Consider another scenario in which my supervisor praised my decision to do the job safely. How would the reaction and resulting stories have differed? Even if the supervisor coached the worker into ways to do the job safely other than simply slowing down, the reinforcement of the perception that safety is important would have been reinforced rather than challenged.

One instance or one story is not usually enough to make a significant difference in the culture. Culture forms over time and is often based on the preponderance of experience and stories. It might take months or years to begin to create the perceptions that make up the first step of the culture cycle. However, if you do want to change the culture, think what stories could help to do so? How could you give workers the experiences that would start positive stories that could change the culture? The better you understand how this process works, the better you can develop a strategy to impact it in positive ways and move it toward improvement.

Case Study: When searching for ways to improve safety and safety culture, many companies administer safety perception surveys to identify areas for improvement. A common category that is probed is management's support for safety. When positive perceptions are identified, many organizations will move on to other categories. However, it is important to understand what is influencing a positive perception within a culture if you would like it to persist.

Recently, I was assessing the culture of a company. The organization is considered one of the best in their industry in safety, yet they still had a few accidents from time to time. While they were proud of their status, they did not want to rest until they had reached and sustained zero accidents. As part of the assessment, we reviewed their safety management system and history of initiatives. We also administered a customized perception survey and conducted many individual and focus group interviews with people from all levels and shifts of the location. Of the different elements, I was interested to find out what had led to the positive perception among the employees about the support for safety by upper level managers.

When I questioned many of the managers and supervisors about their roles, responsibilities, and expectations for safety, the responses were not ideal. Rather than describing their actions, behaviors, and examples they use to demonstrate behavioral integrity, I instead received comments like, "Well, I keep my people from getting injured, make sure they work safe, make sure they report incidents, and address unsafe actions."

Interestingly, I had a different experience when discussing this with groups of employees. The vast majority of the 80 workers (more than four shifts) I spoke to could share with me, almost verbatim, one of three stories that had recently occurred that reinforced the perception that "management really does care about our safety." I had the opportunity to meet the star of one such story.

Harry (an experienced hourly employee) was working at his station operating a piece of machinery. Just as he was shutting the machine down to prepare for a shift change meeting, he noticed two visitors walking toward his area. They were being escorted by an engineer who was relatively new to the site. He noticed that the only individual wearing the required eye protection was the engineer.

Harry walked up to the group, introduced himself, and welcomed them to his area. He informed the visitors about the importance of safety at the site and expressed concern about the lack of proper PPE. Harry let the visitors know that he was concerned for everyone's safety, including visitors to the site. He expressed the importance of eye protection and that it was a requirement in this part of the location. He then asked the visitors if he could show them where the eye protection was stored. After the visitors donned the PPE, he thanked them for their time and wished them a safe visit.

What Harry did not know is that Angela, the assistant site manager, was helping a nearby supervisor and had witnessed most of the conversation. Once the group had departed, Angela approached Harry and asked if he had a moment. She then proceeded to shake Harry's hand and thank him for taking the time to look out for the visitors' safety. She expressed that safety was an important part of her personal values and, moreover, the values for the site. She stated that it meant a lot to her, personally, that Harry took the time to ensure others were safe.

The leaders of this organization were shining examples of those who truly cared for the safety of the employees at their location. However, when I posed the question "What do you believe is creating the positive perception around support for safety?" the group consensus was that it had to do with the number of times safety is mentioned in meetings and the plethora of posters throughout the site. Communication in safety is critical. You often set the priority for something based on how many times you talk about it. If you mention safety once for every 100 times you talk about production, the perception that production is more important than safety is frequently assumed.

When I shared these stories to the leaders during my closing discussion, the managers who were involved in the stories had, for the most part, forgotten about the events. Furthermore, all were unaware of the impact they had on the hourly portion of the culture. The leaders of this site repeatedly communicated their support for safety; what created the value of safety was the cultural reinforcement.

Employees mentioned that when someone expressed the feelings of pressure to take a safety short cut, a fellow employee would provide a story describing a time where a leader showed his or her support for taking the time to follow safe practices. This often comforted the employee and influenced the decision to carry out safe production.

For values to be integrated into the fabric of a culture, they have to be enforced and reinforced at or near the point of decision. Conversations about facts and figures do not necessarily ensure safety values, but stories can be told that will influence decisions on a daily basis. These stories are often the centerpiece of cultures, and your culture is your best reinforcer of values.

Ralph Waldo Emerson once said, "What you do speaks so loud that I cannot hear what you say." What are you doing, specifically, to positively demonstrate behavioral integrity and to reinforce the values of and support for safety? Are people telling your stories, and are they ones your mother would be proud of?"

Questions for Exercise 5

1. What perceptions do workers have about the current safety culture?
2. Are these perceptions accurate or inaccurate?
3. How would you describe the average attitude of safety among workers?
4. Does the organization or individual managers or supervisors actively try to change attitudes?
5. Is safety viewed as a stable value or a changeable priority? How does that differ among levels?
6. Do workers realize the impact of their decisions on the probability of accidents?
7. Do workers make decisions based on safety or is production the overriding consideration?
8. When a worker makes a decision to work safe, what do they expect to happen in the organization?
9. What behaviors do workers regularly exhibit in relation to safety?

EXERCISE 6: THE BRIDGE TO SAFETY EXCELLENCE™

Background Materials

The goal of this book is to help to create a safety culture that can sustain and continuously improve safety performance. The purpose of this section of the book is to
help develop a safety strategy. When most organizations begin to consider how to
develop a more excellent culture, their efforts so far have focused on compliance.
Culture is the second part of safety excellence, and it is important not to overlook
the first part (compliance) as well. The two parts are not redundant nor does the
second part replace the first. They are both equally important, and it is critical that
one precede the other. The diagram in Figure 8 is an analogy of how these two parts
work together to take your organization from its starting place to the desired level
of safety excellence; it is called The Bridge to Safety Excellence.

As organizations begin to mature in safety management, they often recognize
that their culture is the next frontier in achieving safety excellence. Unfortunately,
as work begins on culture, some organizations take emphasis away from more traditional safety approaches. The thought is that if the culture is right, the enforcement
will not be necessary. Unfortunately, this is seldom the case. The compliance part
of safety is essential and does not disappear as the culture matures. The emphasis
will shift from the compliance piece to the cultural piece largely for two reasons:
(1) because culture will begin to have more potential for further gains and (2)
because the compliance piece will have been largely mastered, and there will be
little more improvement to make in that arena.

Rather than thinking of traditional safety as obsolete or simply a tool for
influencing safety culture, think of the two working collaboratively. Traditional
safety ensures and enforces compliance with mandatory conditions and behaviors.
Cultural safety enables and reinforces personal empowerment with discretionary

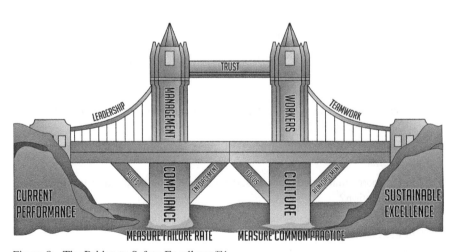

Figure 8   The Bridge to Safety Excellence™.

conditions and behaviors. The choice in safety improvement is not which of these two approaches to use, but how to best use them together. When you move on from basic skills to advanced skills, you do not discard or ignore the basic skills. Rather, you assume them and build on them like you would a foundation.

Traditional safety largely is safety management. It involves managers and safety professionals assessing risks and addressing them through interventions of workplace conditions and practices. It also is an effort to ensure compliance with laws and regulations designed to improve safety. The common tools of traditional safety include rules, procedures, permitting, training, audits, certification, supervision, enforcement, punishment, measurement, incentives, and others. The aim of these efforts is to accomplish the things in safety that are mandatory and prudent. The result of such efforts over the last several decades has been a significant reduction in workplace injuries. Unfortunately, these efforts seldom have eliminated all workplace injuries.

Traditional safety's limitations largely are built into its methodology. Efforts to expand these methods to impact culture have met with limited success. Traditional safety largely is about what management does, what is required, and what the organization officially rewards or punishes. Culture deals with these issues but also is about what workers do, the willingness to go above and beyond the basic requirements, and what the workforce unofficially rewards and punishes. The limitations of traditional safety often are supplemented well by the strengths of cultural safety.

So, what would a model look like that utilizes safety culture to supplement and complete the traditional safety program? One possibility is the model of a bridge that spans from your organization's beginning safety performance to where you ultimately would like to be. The elements necessary to achieve safety excellence basically are conditional and behavioral, so the bridge surface is a layer of conditions that enables the layer of safe behaviors. Where one layer is weak, the other needs to be strong to compensate.

The first pillar of the bridge is based on compliance and is directed by management. It is management's job to know the laws and regulations, to assess the specific risks and address them through the engineering hierarchy of controls. This includes design and engineering of facilities, providing of proper tools, equipment, PPE, environmental controls, and emergency-response capabilities. It also involves the formation of rules and procedures to minimize risks that cannot be engineered or guarded out. Employees must be trained in these rules and procedures, and compliance must be enforced. The quality and completeness of the rules and effective and timely enforcement of them are the braces that make this pillar of the bridge solid.

The second pillar of the bridge is based on culture and involves the entire organizational population. It is the first job of all workers to be in compliance as directed, but the opportunities for further improvement in safety come from going above and beyond compliance. The workers cannot forget compliance or the results will backslide, but they must go beyond it to create Safety Culture Excellence. While the compliance pillar is made strong by rules and enforcement, the cultural pillar is made strong by focus and reinforcement. Managers can enforce, but the culture reinforces. When a safety precaution becomes "the way we do it around here" and workers are willing to intervene with each other to make that happen, the precaution becomes cultural. By learning to focus on improvement targets and reinforcement them among themselves, the workers learn the basic unit of safety improvement.

The supports of the bridge surface are the traditional safety efforts and the cultural safety efforts. The traditional must come first and help to address the basic and required elements of safety. Where the traditional support begins to meet its limitations, the cultural support takes over and addresses the advanced and discretional elements of safety that are not required but critical to safety excellence. Between the two, there is little room for accidents.

The top of this diagram includes additional bracing for strength. The brace that makes management strong and effective is leadership. Simply managing is not necessarily leading. Leading safety means establishing a strategy and carrying it out, communicating it and enforcing it. Leadership is seldom taught to those who are expected to do it until after they get the position. Helping manager learn how to lead safety is critical to Safety Culture Excellence.

The brace that spans between managers and workers is a piece called trust. When trust exists, true cooperation and collaboration can happen. When workers trust managers to have their best interest at heart, they comply and excel. When trust levels are low, it is difficult or impossible to create the interaction necessary for excellence. Trust is compared with a 5-gal bucket that must be filled with an eye-dropper, a few drops at a time, but can be kicked over and spilled all at once. Trust takes a lot of time to build, but it can be destroyed almost instantly. The trust does not have to be perfect to start the journey toward Safety Culture Excellence, but it needs to be sufficient and moving in the right direction. If you are not sure where the trust levels are in your organization, start to ask workers and listen carefully to their responses.

The top brace that strengthens the efforts of the workers is called teamwork. In some organizations, teamwork happens almost spontaneously. In others, it takes a lot of time and work. In either case, it is worth the effort and essential to Safety Culture Excellence. Workers who work together well can accomplish more than those who work alone or distantly. It is difficult for a lone worker to learn and master every aspect of excellent safety performance. If the group is learning and sharing and doing collective problem solving, excellence is attainable, faster, and more fun.

At the bottom of each pillar is the way in which its effectiveness is measured. The compliance pillar's effectiveness metric is basically the traditional lagging indicators of safety. These constitute the failure rate of their efforts: how many injuries per exposure, the severity, lost days, and costs. These metrics are not the only way to measure compliance, but they are strongly traditional and are required by law in most parts of the world. That means they will not go away even if better metrics are used to supplement them. The culture pillar can be best measured by the common practice that grows from the cultural influences. While culture is not common practice, but the common influences that form it, the practice is much easier to see and measure. Theoretically, the better and more consistent the common practice becomes, the lower the failure rate will be. Compliance drives failures to a low level, and culture attacks the gap that remains.

When creating your safety strategy, it is critical that you address both the traditional safety management process as well as planning to create a safety culture that can continuously improve and provide excellent performance. Think of it as working on the basics and advanced issues. They are not the same, and one does not replace the other. They are not redundant, even when used simultaneously. They are simply two STEPS to get your organization to its desired destination.

Questions for Exercise 6

1. Do the leaders of the organization think of safety in two parts: compliance and culture?
2. Has management been successful in getting worker compliance in safety?
3. Does the organization have a good set of safety rules and are they regularly and effectively enforced?
4. Is the safety effort led effectively?
5. Is there trust between managers and workers in safety?
6. Are the workers actively improving the safety culture?
7. Do they have a safety focus and do they reinforce it with each other?
8. Is there a sense of teamwork in the safety efforts?
9. Do you have metrics to show you how you are doing in compliance and culture?

EXERCISE 7: MAKING THE DECISION TO MOVE FORWARD

Background Materials

The decision to of whether or not to move forward with STEPS may not be based exclusively on safety issues. Leaders have other priorities, and the urgency and importance of them may be a consideration in this decision. However, the fact that this methodology can be followed at any speed should also be a consideration. The primary goal of most organizations in safety is to improve, and following these STEPS is a way to structure that improvement and ensure that it is not lost amid other priorities.

The six exercises that preceded this section should have prepared leaders to have a more comprehensive view of where the organization is in its safety progression, what a good safety strategy might entail, which key personnel are good at change and which might be more resistant, and how most organizations move from a compliance to a cultural focus in safety. This comprehensive overview was designed to facilitate a more informed decision. The rest of the information needed is largely an overview of the complete path forward.

The most comprehensive way to map out the path forward is for each leader to read the rest of the book and learn the STEPS. Short of that, here is an overview of remaining STEPS that can create a condensed picture along with the flowchart in Figure 9.

If your leaders decide to proceed with STEPS, here is an overview of the path forward. The path will involve STEPS that can be either skipped (if they have already been accomplished) or addressed one at a time. Each set of STEPS lead to an organizational milestone. The progression is outlined as follows.

1. *Safety strategy.* The next step is a workshop or series of workshops in which the leaders will be asked to further review the issues and make decisions about each of the 11 steps of safety strategy mentioned in Exercise 2.

2. *Assessment.* The organization will be encouraged to a perform an assessment of its current safety culture, which will include measuring perceptions through a survey and interviews, assessing the effectiveness of safety programs and processes, and analyzing accident and/or near-miss data to identify trends and opportunities for improvement.

3. *Clarity.* The next step involves creating a clear vision of what safety is and how it can be improved and sharing the rationale for doing so with the general population. Decisions will be made to either utilize an existing safety team or committee to lead the organization through the remaining STEPS or to form a new one.

4. *Climate.* The organization will focus for a time on each of four STEPS to create a climate that encourages the growth of Safety Culture Excellence.

5. *Chemistry.* The organization will progressively focus on each of eight elements that collectively form a chemistry that encourages the growth of Safety Culture Excellence.

6. *Control.* The team or committee leading the STEPS process will develop the capabilities and skills to identify and address in logical order the conditional and common practice issues that challenge safe performance. The remaining STEPS will be focused on these issues one at a time.

7. *Continuous improvement.* Once the organization is proficient at addressing safety issues one STEP at a time, they will expand their focus to continuous improvement of their process. This will include several STEPS to address issues such as ongoing training and succession planning. Then it will periodically revisit previous STEPS to address issues and to seek opportunities to further improve the cultural issues that promote excellence. Thus, the process will become the standard procedure for how the organization continues to address both safety issues and cultural issues. In short, improvement will become cultural.

All good decisions about organizational change are based on three criteria:

1. The advantages and disadvantages of maintaining the status quo.
2. The advantages and disadvantages of making the change.
3. The exact nature and costs of the change.

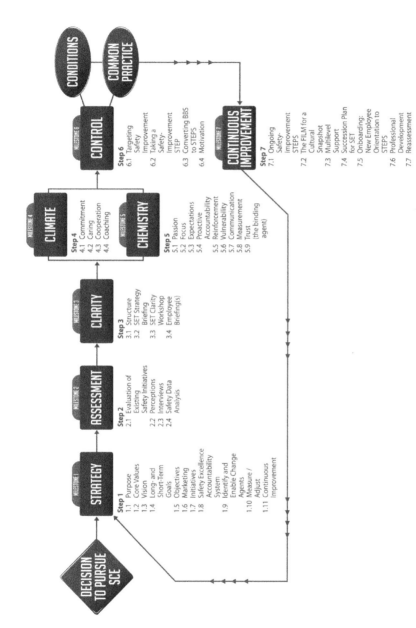

Figure 9 STEPS flowchart.

Questions for Exercise 7

1. Are we satisfied with our current performance in safety?
2. Will our current efforts be sufficient to help us improve?
3. Do we have the kind of safety culture we want?
4. Where will we be in 2–5 years if we continue on our current path?
5. Does the STEPS methodology seem to be a good fit for our organization?
6. Do we want to begin a movement toward improved safety culture?
7. What is our rationale for further improvement in safety?
   (a) Will it increase job satisfaction?
   (b) Could it increase productivity or quality?
   (c) Could it improve our ability to recruit new talent?
   (d) Could it enable us to win contracts with potential clients?
   (e) Does it fit our vision and mission?
   (f) Will it improve our public image?
   (g) Is it a part of the legacy we want for our organization?
8. Can we dedicate resources to this effort?
9. At what pace can we expect to progress?
10. Do we have sufficient internal resources or will we need outside help?
11. Do we have sufficient support among key personnel to ensure success?
12. Can we justify this effort to the owners, board of directors, corporate officers, and so on?
13. Do we have a clear vision of where this path leads and what it will take to move forward?
14. Should we proceed? On what timetable? Who should lead and facilitate the effort?

STRATEGY

> *Business is like war in one respect. If its grand strategy is correct, any number of tactical errors can be made and yet the enterprise proves successful.*
>
> —*General Robert E. Woods*

If you are still reading, you have probably decided either to move forward or at least to explore the possibility of doing so. Always remember that STEPS is not a new program. It has no necessary additional structure and no set timetable. It simply asks you to utilize existing structure wherever possible and to progress systematically through this series of STEPS to ensure that you have the necessary reinforcement to promote excellence and then have the capabilities to get in control of your safety issues by prioritizing and addressing them one at a time.

It has been suggested to us many times that an assessment should be the starting place for the journey to Safety Culture Excellence^{SM} rather than developing a safety strategy. In our experience, when you begin with the assessment, your strategy can become simply a plan to address your weaknesses rather than a true strategy. It is like planning your life based on a visit to your doctor's office. A strategy should give direction and meaning to everything else you do in safety. If you have this in place before you assess your current status, you tend to move forward toward your goal rather than from side to side addressing your perceived issues reactively. One of the challenges of Safety Culture Excellence is to move from strictly reactive to proactive efforts. Developing a safety strategy is the first and most important step toward doing so.

Goals: To move from avoiding failure to achieving success

To include excellence in the safety vocabulary

To align all safety activities around an overarching strategy

To illicit extra effort by defining the rationale of safety

To provide a clear and repeatable direction toward success

To align and motivate workplace behaviors to accomplish the strategic goals

---

STEPS to Safety Culture Excellence^{SM}, First Edition. Terry L. Mathis and Shawn M. Galloway.
© 2013 John Wiley & Sons, Inc. Published 2013 by John Wiley & Sons, Inc.

Methods: A leadership training and workshop or multiple workshops to develop a Safety Strategy

STEPS: 1.1 Purpose

1.2 Core Values

1.3 Vision

1.4 Long- and Short-Term Goals

1.5 Objectives

1.6 Marketing

1.7 Initiatives

1.8 Safety Excellence Accountability System

1.9 Identify and Enable Change Agents

1.10 Measure/Adjust

1.11 Continuous Improvement

Now that leaders have decided to move forward and have the big picture of what a safety strategy entails, they can participate in a workshop or series of workshops to fill in the details of a customized strategy for the organization. The following sections are dedicated to each of the 11 elements of a safety strategy. If it has been too long since the first training/workshop in which the leaders reviewed the materials mentioned previously, it may be wise to review each of the 11 elements on the diagram in more detail from the previous section. Do not be concerned if you do not complete every detail of the strategy in this session. Strategies are living things that need to grow and change as thinking and issues change. If leaders cannot flesh in their strategies in a reasonable time or seem to be stalled in their thinking, it is usually best to skip over that section and move on through the others. You can revisit the skipped session at the end, but there is still no pressure to develop the strategy during this session. It is better to get it right than to get it right now! Remember also that you are going to go through each of the STEPS and that strategies might become clearer during that activity.

Icebreaker Activity: To begin the discussion about strategy, get leaders to think about a strategy that may be more familiar than safety strategy. The example we are using here involves a strategy to dominate market share for a product or service. (If this example is too foreign to what the organization does or if you have a better example at hand, please use another one for this activity.) If this was the challenge, the leaders might follow a model like the one in Figure 1.1. In each of these STEPS ask the leaders for their answers to the questions and how these answers could help develop a strategy.

The STEPS of this process are as follows:

1. Make the business case for the product or service. What is its function? Why is it needed? Who will buy it? What other products or services are on the market and how do they compare? All these questions would be answered, and a statement of purpose would be developed. It would explain not only the product or service but also how the organization would benefit by offering it.

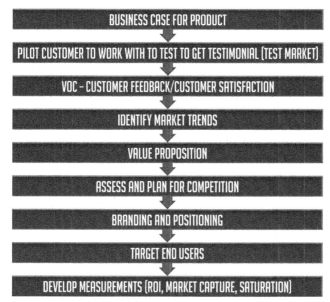

Figure 1.1   Ice breaker activity: Example strategy to dominate the market.

2. A pilot customer would be identified based on the profile of the product, and this customer would be asked to test the product and endorse it. During the course of this process, the product might be found to have faults or weaknesses or there might simply be opportunities to make it better and more suited for purpose.

3. This would continue until the customer either accepted or rejected the product. The voice of the customer (VOC) would be heard and taken into consideration in both the design and the marketing approach for the product.

4. The organization would conduct, or contract for, market research. This research would establish the product's potential place, but this would also look for trends in the product type. Is this a product with a growing demand, a steady demand, or a diminishing demand? Is the price of similar products moving in a definable direction?

5. An analysis would be made to determine how valuable it would be for the organization to make and market this product. What are the profit margins and the volume potential? How long would it take to begin and how much would it cost to begin? What would profits look like over the projected life of the product?

6. Who is the competition and who could become the competition? How would they compete and what impact would competition have on the profits and life cycle of the product?

7. How will the organization brand the product? What is its name, logo, who is the spokesperson? How will the product be viewed by the potential buyers and how can that best be managed?

8. Who, exactly, are the potential buyers of this product? What are their demographics: income level, neighborhoods, work places, what do they read or watch, and how can they best be reached?

9. The organization would decide how to measure the key indicators of the success of this product launch and ongoing life span. What is the return on investment (ROI)? What is the market share and the rate of market capture? What is the percentage of market saturation?

All these STEPS and questions would be a part of the planning and consideration for such a strategy. How many of these apply to a safety strategy? Can we do significantly less in developing a safety strategy than we do in a new product strategy and expect the same probability and degree of success? The sad truth is that most organizations run other aspects of business with much greater attention to detail than they do with safety. This is the reason that many excellent organizations have less-than-excellent safety performance.

With this level of detail and this comparison of safety strategy to other strategy, the leaders of the organization should begin to consider the elements of a safety strategy. Refer back to the STEPS in Figure 4 in "Making the Decision to Pursue Safety Culture Excellence" that are relisted as the headings of each of the sections in the workshop.

Case Study: Recently, a plant manager led an all-hands meeting with supervisors. He did so with the intent of discovering what was contributing to recent injuries. He also wanted to understand if the supervisors were helping or hindering the efforts to improve. Unexpectedly, one of the supervisors asked the question, "Could you help me understand what the strategic direction in safety is?" The plant manager responded honestly, "Well, that is actually a good question. I don't believe we have a clear direction in safety."

The manager looked around the room and asked for confirmation from his department leaders. The body language uncomfortably indicated agreement. The supervisor then politely responded in a hushed tone with a very profound question. He bravely inquired, "Sir, if you don't know what you want us to do, how are we supposed to know? And, how are we supposed to act accordingly?" This supervisor expressed a concern shared by many first-line leaders.

Many well-intentioned executives believe that their strategic safety direction has been successfully communicated. The sad truth is they are often wrong. Irish playwright, George Bernard Shaw, once said, "The single biggest problem in communication is the illusion that it has taken place." If there is no clear, memorable, and repeatable direction, can we really expect people to be working in unison toward the same goals?

Filling in the Details of the Safety Strategy: The remainder of this workshop consists of 11 exercises, each of which is designed to fill in the details of the 11 parts of an effective safety strategy. As you proceed, remember the icebreaker activity and do not fall into the mindset that safety is simpler than other business goals, or that a strategy of getting better by some percent or simply starting a new initiative is really adequate to drive excellence. Encourage the participants to use this workshop to do some deep analysis of how and where and why safety is critical to the

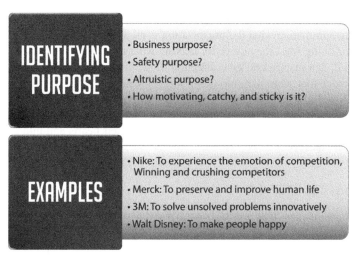

Figure 1.2   Identifying purpose.

organization and what it is really about. We would like you to come away from this workshop with not only a strategy but also a group of leaders who really "get" what safety is about, why it is critical, and how to make it excellent.

STEP 1.1   PURPOSE

Background Materials

Organizational leaders should be challenged to do some soul-searching and to really determine why they choose to work toward safety excellence. There is seldom one single answer to this question. Use the diagram in Figure 1.2 to center this discussion.

- *Business purpose.* There is often a business reason or reasons for improving safety. Some may want to reduce the costs and/or other negative effects of accidents. Others may see competitive advantage such as landing contracts with client companies who demand safety excellence as a condition of doing business. Others may consider safety failure as a major distraction to production or quality.

- *Safety purpose.* Safety professionals may have their own rationale for beginning new initiatives as a part of their overall strategies of managing safety for the organization. They may be aiming for greater participation in safety efforts by nonsafety personnel or better understanding of risks.

- *Altruistic purpose.* In addition to business and safety reasons, many organizations want to improve safety simply because they care about people and want to reduce pain and suffering. For many, the reality of workplace injuries is personal. If you have had a friend or loved one injured on the job, you may simply want to spare others the experience of such misfortune.

Many organizations have a statement of organizational purpose or vision. These are designed to summarize what the organization is all about and what they see as their purpose for doing what they do or for making what they make. These statements, if created well, can inspire and give direction to the employees. They serve as a constant reminder that what they are doing has a greater purpose than the task of the moment. A well-crafted statement of purpose can provide the same benefits to your safety strategy. It can also be the beginning of sharing the rationale for this effort with the other members of your organization to get their hearts and minds involved in helping to succeed.

Questions for STEP 1.1

1. How would excellent safety benefit your business?
2. How would excellent safety benefit the safety department?
3. What altruistic, humanitarian goals of the organization would excellent safety accomplish?
4. Can we turn these three purpose statements into a purpose or vision statement?
5. Would the members of our organization respond well to such a statement?

STEP 1.2   CORE VALUES

Background Materials

In most of the examples our clients have given where they have made initial statements of the value of safety, these statements tend to reflect situational rather than core values. A situational value is a value that prompts an employee to act a certain way in a certain situation. Statements about the priority of safety versus productivity are good examples of situational values. They basically tell workers, "When you must choose to take a risk to get the job done more quickly, this is how we want you to make that decision."

Core values, on the other hand, are universal and transcend the limitations of specific situations. They might apply to specific situations, but they are not limited to them. Values are created when perceptions and behaviors are reinforced continuously, at or near the point of decisions. Figure 1.3 has some examples of generic organizational core values. These are principles upon which the organization wants to base its business decisions and direction. It is the moral and philosophical compass of the organization. Such a statement of core values in safety is rare, but it is extremely valuable in achieving safety excellence.

Beware of statements that are exaggerated or simply unattainable. Many organizations have stated that "safety is job 1" or "safety is our first priority" or that "nothing is important enough to put our employees at risk" only to have it backfire and actually hurt their safety excellence efforts. The reality for most organizations is that they are in business to make a product or to provide a service and that doing

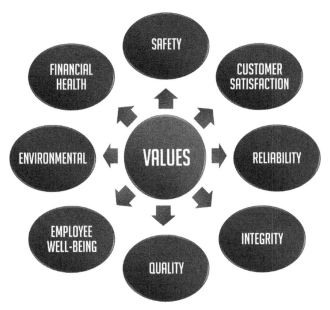

Figure 1.3   Generic organizational core values.

so involves some inherent risks. Conscious organizations do their best to manage those risks and to minimize the probability of workers getting injured on the job. In our experience, no organization has ever created a perfectly safe work place nor trained and managed the perfectly safe workforce. Even if they did, their workers would still be at risk at home, in public, and on the streets and highways. Safety perfection is not realistic, but safety excellence is both attainable and attained already by some organizations. Statements of safety core values should be centered around organizational best efforts, not perfection.

Some examples of safety core values might include:

- *Honesty.* Making sure to be completely open and honest in all safety discussions and reporting
- *Consideration.* Caring about the safety of others and taking action to protect them when necessary
- *Service.* Being willing to do one's part in any safety activity or program
- *Resources.* Making sure that everyone has needed equipment, assistance, and information to be safe
- *Example.* Every employee should consider themselves a role model for safety and always be a good safety example for their fellow workers
- *Compliance.* We will do our very best to know and comply with every law and regulation for safety that applies to our industry
- *Excellence.* We will never consider any number or kind of accident as inevitable or nonpreventable and will continuously strive to become more perfect in our safety performance

Testing Core Values

Another way to think about how you envision and strategically define your safety core values is simply to remember that if you say it, your employees will test it. Ask yourself, can we pass the test? Can we actually do what we say we will do and what we imply? If you say "safety first," does it mean that you will always shut the plant to fix a safety problem? Does it mean that production will always be allowed to suffer if it impacts safety? Does it mean that leaders will always consider safety first before profits or competitive advantage or their own careers? Basically, can your performance reinforce your statements? If not, the statement will likely do more harm than good! Before your employees get a chance to test your core values, test them yourselves and only adopt and communicate the ones that pass your test.

Questions for STEP 1.2

1. What core values for safety are already identified in the organization?
2. Have we identified core values in other areas that apply to safety?
3. What basic qualities would we like to instill in our safety efforts?
4. What are some things we do not want people doing in the name of safety?
5. When we mention safety, what qualities and principles would we like people to automatically think of?
6. What is the ideal example we, the leaders, could set to direct the efforts?
7. Can we build the concept of excellence into our core values in a way that helps everyone understand it better?
8. Can we brainstorm a list of core values that meet our organizational needs now?

STEP 1.3   VISION

Background Materials

Leaders should begin to develop a mental picture that can be perfected and shared with the organization of what can be achieved in terms of safety excellence. This vision should begin with the idea of what safety excellence looks like. Again, this is not a one-dimensional view and may have several different facets. The vision may include what the organization is doing, how communication is happening, what workers are focused on, and so on. It is important to envision more than simply results based on lagging indicators. For example, do not simply say that safety success looks like zero accidents. That is the result, not the effort it takes to produce the result. Even if you reach zero accidents, what will keep you there?

Rather than trying to describe your whole world, try to create a vision that will inspire and direct workers to help you achieve more excellent performance. Hit the high points and describe the overall effort rather than the minute details of it. How will efforts be more focused and directed? How will everyone work more

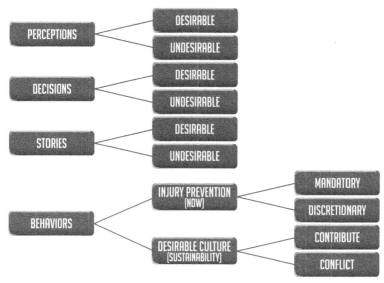

Figure 1.4   Targets for visioning.

collaboratively? How much clearer will everyone's role be? What new ways will we have to measure success rather than just accident rates? How will safety performance be more excellent and what will that mean to the organization and each member of it?

Visioning, if done correctly, creates focus. In Figure 1.4, there is a chart to help you consider what targets you should have for your visioning exercise. There are four targets that are common when improving a safety culture: changing perceptions, changing decisions, changing stories, or changing behaviors. The vision can involve any or all of these targets as needed.

The final analysis of a vision is whether or not it describes a state or place that is appealing enough to motivate effort to get there. Does the vision create desire and inspire a longing for something better or dissatisfaction with the current condition or both? Think about the inspiring visions of famous leaders that have moved peoples and nations to great action: Franklin Roosevelt's vision of what it would take to win the war, or John Kennedy's vision of sending a man to the moon, or Martin Luther King's vision of overcoming racial prejudice. All of these visions motivated people to action and that action changed the current condition. Does your safety vision pass the test? If you are not sure, ask your employees.

Questions for STEP 1.3

(The visioning exercise questions are also found in Figures 1.4 and 1.5.)

1. What does excellence look like in 5 years?
2. What will it take to make this happen?

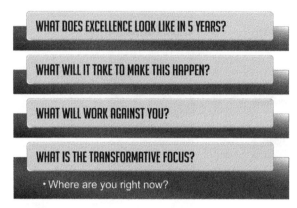

Figure 1.5   The vision of excellence.

3. What would you see people at all levels doing that would indicate that you have achieved excellence?

4. What will work against you?

5. What is the transformational focus?

6. Where are you right now?

7. What perceptions of safety should your vision create?

8. What kind of decisions about safety should people make if they follow the vision?

9. What kinds of stories should begin to circulate that will reinforce the perceptions created by the vision?

10. What behaviors will people feel appropriate doing if they are following the vision?

11. Will this vision and these perceptions, decisions, stories, and behaviors help create an excellent safety culture?

STEP 1.4   LONG- AND SHORT-TERM GOALS

Background Materials

As you begin the STEPS process, you will be constantly targeting certain improvements and working on them. For now, you should consider the big picture of what you are trying to accomplish. Of course you want to improve safety results, but what are the techniques and methods you will use to accomplish the improvement in results? What can you accomplish now and what can you accomplish over a longer period of time? Where would you like to be in 6 months or 5 years?

To reach a level of excellence in safety performance, it is desirable to set positive rather than negative goals: for example, a goal to get everyone to take a precaution versus getting everyone to avoid a danger. On the surface, these seem very similar, but the differences between them become more and more profound as the organizational performance begins to improve. The overall difference is between trying to achieve success versus avoiding failure.

Positive goals can be reached by positive reinforcement, which is a powerful tool to build excellent performance as well as strong relationships. Negative goals often involve blame and punishment designed to stop risk taking. The tools to effectively stop human behavior often damage relationships and attitudes and create an environment more akin to a police state. We have never seen an organization blamed or punished into excellence. Excellence is a function of performance, and the best performance is a matter of building on strengths more than eliminating weaknesses. Strong cultures are built on strong relationships. A strategy that encourages using relationship-damaging tools is not a viable strategy.

Good short-term goals should match the overall strategy and be aimed at moving the workforce toward better safety performance. They might include:

- Making every employee aware of the new safety strategy and answering questions about it
- Creating a mindset of focusing on a single safety problem and solving it before moving on
- Creating and communicating metrics that show progress toward the single goals
- Enhancing safety communications to keep everyone focused on targeted improvements.

Long-term goals should help complete the aim of the overall strategy by creating culture and other structure that can sustain excellent performance. They might include:

- Developing better safety metrics to understand how efforts turn into results
- Improving employee interactions through training and models of communication
- Continuously increasing opportunities for employees to get actively involved in safety excellence activities
- Have employees think of managers and supervisors as safety coaches who can help them perform better and safer rather than safety cops who are strictly out to enforce the rules and catch the violators.

Remember that there will be many more specific goals set to improve safety as you progress with STEPS. These initial goals should be designed to help start your safety strategy to becoming the way everyone thinks and acts in safety. These initial goals should prepare everyone in the organization to begin the journey toward Safety Culture Excellence that will see many goals set and accomplished and celebrated. If the goals are overly broad or unrealistic, they will not accomplish their purpose. In fact, they can actually create the perception of a false start or even a failure of the process. If they are too simple, unchallenging, or not worthwhile, they might not light the spark needed to motivate progress.

Questions for STEP 1.4

1. What short-term goals will get quick wins and motivate our process?
2. What short-term goals will quickly build interest and participation in the process?
3. Which aspects of our vision and core values can be improved quickly?
4. Which short-term goals have obvious, automatic long-term goals?
5. What long-term goals will motivate short-term effort?
6. How much can we realistically work on at once?
7. Do we have enough goals to motivate and not too many to discourage?

STEP 1.5   OBJECTIVES

Background Materials

If you reach the goals, what will you accomplish and how will this further your safety strategy? This step is a reality check for the goals you have set. Analyze each of them and see if they are really worth accomplishing and are fully aligned with the direction you are developing for your safety strategy. Where does each goal fit into the framework of your strategy, and are there any elements of the strategy that have no goals to direct effort toward them? If you accomplish this goal, what will be the impact on the safety culture? Will it be more capable, better informed, and more focused? Asking such questions can help to redefine or sometimes eliminate goals before you waste time and effort working on the wrong, or poorly defined, items. Basically, the goal is what you are trying to do, and the objective is what you are trying to become.

Questions for STEP 1.5

For each goal, ask:

1. If we reach this goal, what objective will we have accomplished?
2. Is this truly our objective?
3. If we reach this goal, will our organization become more excellent in safety? (Do not forget that some goals are intermediate: they move you closer but not all the way to the ultimate goal.)
4. Will this goal accomplish something permanent, or will this issue need to be reviewed periodically to see if it needs further work?

STEP 1.6   MARKETING

Background Materials

The STEPS process involves a complete look at how you market safety to the whole organization. For now, you should think how you will begin to market the safety strategy. Why do we use the term "market?" Because we believe that your employees are the customers of your safety strategy and that they must "buy in" if you are going to be ultimately successful in reaching excellence. Before you panic and declare yourself "not a salesperson," consider that we are talking about marketing and not sales. We do not want to sell someone something that is not ultimately what they want and need. We also do not want to resort to force (command-and-control management) to make the strategy work, since force can damage our long-term goals and the relationships needed for true excellence. Marketing is getting the right message out to create interest in the product among the right people and to reinforce their decision to remain interested in the product. For now, we are not talking about how to get the message out. That is communication. We are talking about what the message should be.

As you consider how to get the buy-in you want for your new safety strategy, focus on these four important aspects of marketing in Figure 1.6 that can guide your plans. Remember that your marketing strategy is changeable and that you will be asked to revisit safety marketing as a STEP later in the process. You can expand your thinking and perfect your marketing later, but now you will need to market your strategy to get the ball rolling, so decide the best way to do so. These four important aspects are:

1. *Branding.* How will employees recognize your safety strategy and how will they picture or envision it? Products and services often have "brands" that include names and/or logos that label and identify them to customers and potential customers. Just the term "safety" has a certain brand identity, but it may not be the one you most want. Many safety programs and processes have branded themselves with names and logos designed to appeal to the targeted customers. If you want emotional involvement, you can use inspirational names and logos. If you want participation, you can use team names and logos.

Figure 1.6   Four vital elements of Marketing Strategy™.

Sometimes the safety strategy will almost name itself from the wording of your goals and objectives. Sometimes it is best to let the employees name the process to elicit their involvement. Whether you utilize names and logos, mission of vision statements, or simply describe the importance of the effort, keep in mind how you would like everyone to think of and picture your safety strategy. What would you like them to associate it with and what kind of image would you like it to have?

2. *Positioning.* The term "positioning" is relatively new to marketing. It first appeared in 1969 in a book by the name *Positioning: The Battle for Your Mind*. The authors, Jack Ries and Jack Trout, defined positioning in their paperback edition on page 19 as ". . . an organized system for finding a window in the mind. It is based on the concept that communication can only take place at the right time and under the right circumstances." So when is the right time to communicate safety, and what are the right circumstances? The answer to these questions almost necessarily includes explaining the rationale for wanting to improve safety. Where do leaders place safety in the minds of employees in relation to other values and priorities? How do they make the case for spending energy and resources on this effort? All these considerations will equal the positioning of the new safety strategy. If done properly, this aspect of the marketing plan will define when and under what circumstances the new safety strategy should be communicated and exactly what the message should contain.

3. *Voice of the customer (VOC).* Products and services can fail when they are designed without input from the people who are expected to be the customers. Almost all good marketing research includes asking people what they want, what they like and do not like, and getting a reality check on the product. A safety strategy can meet with the same fate if the customers (employees) have no voice in the design or features. Really excellent marketing research tests the VOC before, during, and after the design. A design workshop could begin with one or more focus group meetings in which employees are asked what they would like to see different in their safety programs. When the strategy starts to take shape, they could be asked for input. Then, when the first draft of the final safety strategy is ready, ask again. It is almost always easier to make modification early in the process than after new programs and processes are in place.

4. *Reinforcing the buying decision.* As you finalize the plan to market your new safety strategy, you should ask the question, "What will make our employees glad they bought into this new strategy?" A part of marketing is making sure buyers do not wish they had not bought just after the sale. What could make employees feel that they made a good decision to support the safety strategy? What if they were thanked publicly for their support? What if they get progress reports regularly on how the new strategy is rolling out? What if they see some quick wins and it looks like the new strategy is producing results? What else specifically in your organization could reinforce the buying decision?

Questions for STEP 1.6

1. What kind of name and/or logo would brand your safety efforts accurately and effectively?

2. What kind of name and/or logo would your people buy into and get excited about?

3. Where would you like to place safety in the minds of your people in relation to other priorities or values?

4. What associations would you like people to make when they think of your safety initiative?

5. How can you emphasize that safety is listening to and meeting the needs of its customers?

6. Can you answer the what's in it for me (WIIFM) question in your marketing message?

7. How can you make people proud of their decision to support safety?

STEP 1.7   INITIATIVES

Background Materials

Initiatives are projects, programs, and other efforts to help accomplish the goals and objectives. It is important to have clear goals and objectives to align initiatives to directly accomplish them. Too many initiatives are aimed at ambiguous targets such as "improving safety" or "increasing awareness" and not at specific, targeted goals. When you define clearly what you want to accomplish and the STEPS to get there, it is much easier to develop new initiatives or to choose existing ones to directly target results.

In addition to considering new safety-improvement initiatives, the organization should begin to analyze and align existing initiatives with the newly formulated safety strategy. Each program or initiative should be labeled as to what its unique contribution is to achieving the goals and aiding the methodologies of the strategy. Once they are aligned, they can also be prioritized by the impact they have had or potentially could have on achieving safety excellence. Resources can be allocated to each according to their needs and potential contribution.

Any safety programs or initiatives that have key performance indicators (KPIs) can also be evaluated to see if the process or initiative KPIs can contribute or become a part of the newly defined safety metrics. As the organization begins to move from strictly managing safety with lagging indicators toward a more robust and eventually balanced-scorecard approach, the process metrics can often be valuable to the overall strategy as well as to the individual initiative for which they were designed.

Any initiative that is not aligned with or contributing to the new strategy can be eliminated, reduced, or realigned. If you are going to replace an initiative with a new one to achieve your safety strategy or to better utilize STEPS, first consider if the existing initiative is functional and contributing to safety results. Functional and integral initiatives should be continued and phased out as new replacements come up to speed rather than scrapped and replaced over time. The gap between stopping the old initiative and starting the new one can produce undesired results. It can also demoralize those participating in the old initiative and maximize the perception of change. The bigger the change is perceived to be, the more resistance it can create. Keeping the perception of change minimized is almost always a worthwhile goal as you progress toward Safety Culture Excellence.

Reviewing initiatives to align them with your new safety strategy can also be an opportunity to breathe new life into old initiatives. If you have a safety initiative that has been around for a long time and is growing stale or routine, consider bringing those involved together and challenging them to realign their program or initiative to become a contributing part of the new strategy. Offer some training and assistance, if needed, and allow the old horse to get in shape for a new race. One of the most common problems with safety initiatives as they mature is that the people involved become oriented to keeping the process alive and going through the motions, and they can lose sight of the original goal of improving safety. Reigniting the "results" orientation can replace the "process" orientation with a much more energized initiative with a very specific sense of purpose.

Questions for STEP 1.7

1. What existing safety programs, processes, or initiatives are still active in the organization?
2. How exactly do each of them contribute to the new safety strategy?
3. Do we have some initiatives that do not add value or contribute to the safety strategy?
4. Could the nonvalue programs be realigned to make them valuable?
5. What would be the potential negative impact of discontinuing these programs?
6. If we discontinue a program, how can we utilize the people involved and not demotivate them?
7. Are the contributing programs functioning well: results oriented versus process oriented?
8. Do we need other initiatives to help us accomplish our safety strategy?

STEP 1.8   SAFETY EXCELLENCE ACCOUNTABILITY SYSTEM

Background Materials

If your organization has job descriptions or other attempts to define personal responsibilities for safety, they will likely need to be reviewed and revised to meet the needs of the new safety strategy. You may decide that what you have is good enough for now and visit it later as you continue through the remaining STEPS. However, if you do not have job descriptions (or if they do not include safety), you should consider at least getting started. We recommend a format we call safety roles, responsibilities, and results (SRRRs). This format defines the three most important elements of human performance in safety: what a person should be (the roles they perform), specifically what they should do to fulfill those roles (responsibilities), and the outcome expected if they perform their roles and fulfill their responsibilities (results). Defining specifically these three important aspects of performance avoids many of the common problems with job descriptions.

Job descriptions tend to be a simple list of tasks or responsibilities to be performed. While these are critically important, they tend to also define what people do not have to do and excuse them from producing results as long as they fulfill their tasks. SRRRs are a more holistic approach to describing what is expected and provides a more detailed way to determine accountability. The three "Rs" ensure that it is clear what everyone in every position should be, do, and produce.

Obviously, SRRRs will differ from one position to another. Contributions to accomplishing the goals using the safety strategy will be divided into different levels and types of contributions. While each player will have a different way in which to contribute, the sum of the contributions will constitute the whole job of success. Excellence in a team is not necessarily a team effort. It is all right to have superstars, and it is unrealistic to think that all contributions will be exactly equal. However, the magnitude of work is often offset by the need for specific work that some can do better than others. The importance is that the whole job of safety excellence gets done by the team and that everyone contributes in their own most effective and efficient way.

The development and wording of SRRRs may be ongoing in many organizations. Needs change and opportunities arise that may mandate the modification of SRRRs over time. However, there are two critical elements to making them do their best jobs:

1. They must be incorporated into performance appraisals or evaluations quickly. It is not enough to have clear expectations; these expectations must be reinforced. It is also important to establish that safety excellence is an organizational priority and an integral part of everyone's job. This cannot happen if there are no consequences for either meeting or failing to meet expectations. Accountability means that an accounting must be made of performance against expectations. However, remember that we want hearts and minds involved also and that punitive systems are not usually effective ways to accomplish this. Good performance should be praised, and poor or unacceptable performance should be corrected or rechanneled. This can be done as a matter of setting expectations without the need for artificial punishment for minor offenders.

2. They must be evaluated to make sure that someone is assigned to each important element of the safety strategy. Someone or some group in the organization must have a part of his, her, or their roles, responsibilities, and results (RRRs) to compare the group of RRRs with the safety strategy and to determine that

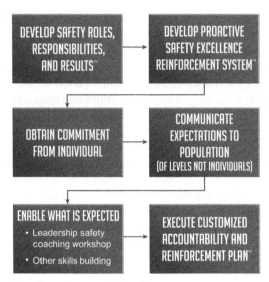

Figure 1.7   Safety Excellence Accountability System.

no important RRR is not assigned to someone. As the process progresses, it sometimes becomes evident that not enough people are assigned to some tasks, or there are too many assigned to others. If this is discovered, changes can be made to individual or positional RRRs to compensate. Many organizations assign the oversight of SRRRs to their Safety Excellence Team (SET) (which will be discussed in the section on clarity).

The process of implementing SRRRs within the organization can best be accomplished by following the process in Figure 1.7. Simply developing SRRRs and handing them to individuals will result in a very slow startup if not a total failure. The way in which you implement SRRRs will be as important as how well and thoroughly you write them. The implementation should proceed as follows:

1. *Develop the SRRRs.* The first step of this methodology is to collaboratively outline the top five RRRs expected of someone in safety. Collaboration is important for ownership. It is more difficult to hold others accountable for what you feel are important responsibilities. Typically, this begins with involvement of the different levels to outline what excellence would look like in the behaviors of individuals of the many major levels in an organization. What would you see them doing or saying that lets you know they are a great leader, very supportive of safety, and their performance certainly will impact results with their reports or peers.

2. *Develop a Proactive Safety Excellence Reinforcement System™.* Like this book outlines, wishing and asking for results are not effective approaches. There must be a strategy and methodological approach to holding someone accountable for key responsibilities. Like safety, there are two sides to accountability: proactive and reactive. Also like safety, there needs to be a balance of

consequences for desirable performance and undesirable performance. This balance of consequences has a role in both the proactive and the reactive sides of accountability. What is in place to recognize what someone is doing or to address what they are not doing, before checking if the results were impacted? The authors believe that this is the true purpose of accountability. However, most organizations better manage the reactive side of accountability, and they focus more on what someone did not do and try to hold them accountable. This typically leads to a poor perception of the approaches and even terminology of accountability. What is your plan? Who will carry it out? How will they reinforce and react? How often will they meet (generally one-on-one monthly) to reinforce the performance necessary and to provide feedback postresults?

3. *Obtain commitment from the individual.* While it is important to first develop your accountability reinforcement system, consider not deploying it until the people involved have provided you their individual, private commitment to focusing on the items necessary. This is where many sites have taken the list created in a previous step in this model and jointly (direct supervisor and direct report) chosen the top three to five items to focus on and master. A long list of responsibilities is very difficult to focus on, and we would argue more than five is not a clear focus for performance accountability. Once the top three to five RRRs are mutually chosen, it is important that written commitment is obtained to focus on and improve in those areas.

4. *Communicate expectations to the population.* To help increase the effectiveness of your approach, positive peer and group pressure can be leveraged to strengthen the sense of self-accountability. While we rarely communicate to a group the individuals' roles and responsibilities, we do encourage the communication of what others should experience (results) if the individuals are being the type of leaders needed to experience transformation. This is also important because one of the mechanisms to determine improvement or consistency of the key responsibilities is the feedback of those that should be experiencing the behaviors of the individual being coached. Are they having more, less, or the same types of experiences when working with these individuals?

5. *Enable what is expected.* Identifying what someone needs to do to contribute to desirable results will only have an impact if the individual has the capabilities to perform in the newly discovered desirable ways. If one key responsibility is that the individual needs to lead better, more participative safety meetings or talks, can they do so? Do they have the necessary speaking platform skills? Do they know where to get topics or how to develop materials? If you expect your leaders to become better performance coaches and evolve from being a compliance police, do they know how to coach for performance? Do they have the necessary skills to deeply understand what motivates performance and how to influence it?

6. *Execute the Customized Accountability and Reinforcement Plan™.* It is vital that a plan first be developed, RRR's customized commitment obtained and

competencies developed and plans to reinforce what is expected, before we can begin to expect improvement in performance. Organizations need not make this a complicated approach; in practice, it has been surprisingly simple to develop and deploy. The biggest challenge is staying the course. If we, as the leaders of this culture improvement, desire for executives to hold managers accountable, managers/supervisors, and supervisors/employees, what is your plan to manage all of this, and what is your sphere of influence? Most organizations effective with this approach begin at the highest levels and cascade downward as each level demonstrates improvement, ultimately pushing down to first-line supervisors or employees. If executives are not demonstrating the ability to focus and coach managers, we cannot expect that supervisors or employees will improve. The old adage is always true; people pay attention to what their immediate boss pays attention to.

Questions for STEP 1.8

1. Do we have adequate job descriptions or roles and responsibilities for safety accountability?
2. How clear are our expectations for safety?
3. How well do our people know what is expected of them in safety?
4. How would we rate our current performance in safety accountability?
5. Do we need to address this issue in our strategy or can we revisit it later?
6. If we decide to revisit it, when will we do that and how will we remind ourselves?
7. If we need to address it now, should we go for a finished product or just a starting place?
8. If we need to address it now, who should be on the team to accomplish the task?
9. Should we begin by asking everyone to submit what they think SRRRs should be for their positions?

STEP 1.9   IDENTIFY AND ENABLE CHANGE AGENTS

Background Materials

Meaningful change seldom happens in an organization unless specific people are selected, designated, and enabled to be agents of the change. The selection of the right kind of people is critical. Some people have influence over others and some do not. The kind of influence they have is also important. If the influence simply comes from position or rank, the change will tend to be dictated rather than engaging. The main principle of cultural change is that the change should ideally come from within rather than from without the culture. This means that "champions" or "mavens" should be carefully selected from within areas of the culture or from the major subcultures to represent the change from within. Ideally, these will be respected and influential members of their respective groups within the culture.

Once selected, these change agents need to be designated as such. Everyone in the group should know who their group's change agent is and consider that person to be the "go-to" person when there are questions or suggestions regarding the change. Change should not be covert. No one should feel they are being manipulated or secretly observed. The real goal of change agents is to make everyone else an informal change agent also. The designated change agent's job is to recruit, enlist, and convert but not to entrap or fool. Never forget that you are trying to change a culture and that everyone within the organization is a member of that culture. The overlying goal of the change is to enlist willing cooperation, not grudging compliance.

When selected and designated, the change agents should be empowered to fulfill their assignment. This means that they should know clearly not only the roadmap of the change that is planned but also the rationale of each step and milestone along the way. These people will be channels of communication and information, but they should view themselves more as sales or marketing people. Their job of informing is a necessary part of getting the buy-in necessary to make the change possible and sustainable. The change agents are asking all their fellow employees to join a new team that is going to win the Safety Culture Excellence title. Winning teams do not just have players who know the plays; they have a team spirit centered on the strategy (or game plan), and their hearts are in the game.

When you begin the formal STEPS process, you will be encouraged to form a SET to steer the process. These team members will be the primary change agents, but that does not mean that you cannot enlist others at the workgroup or department level. As you formulate your safety strategy, consider how your organization is structured and envision what kind of change mechanism will best help you to succeed. How many people, at what levels, doing which tasks, and conveying which information, will make the change happen effectively and efficiently?

Many organizations want to designate full-time positions to function as change agents to get the dedication and effort necessary to be successful. While this idea can have some merit, do not forget that the overall goal is to change the culture and that, in some organizations, as soon as a person gets another assignment, they are no longer considered a part of the original work group. If the change is truly to come from within and if it is to become a common practice (the way we do things around here), that may be best accomplished by part-time volunteers who continue to do their regular jobs and to remain a part of their workgroups. Remember the example of Wikipedia, the strictly unpaid, volunteer group who wrote an online encyclopedia that virtually put the old encyclopedia companies out of business. Those old companies had a full-time, paid staff of professional writers and experts and an existing business and clientele, but they were outperformed by a group of volunteers who formed a new culture in which everyone contributed and benefited. What if you can do the same thing with your safety culture?

Exercise: When you consider making an organizational change or improvement, you will encounter different responses from various persons. In our experience, the people in an organization can be grouped by their response to change.

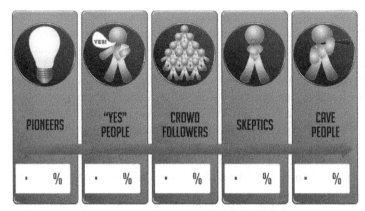

Figure 1.8   The five types of people in Organizational Change™.

Every organization is different and unique, but all tend to have five groups of people who react to change in predictable ways (see Figure 1.8). Look at this figure and estimate what percent of your organization's population falls into these groups, or use this exercise in a group discussion as you develop and roll out your strategies for Safety Culture Excellence.

- *Pioneers.* These are people who like and embrace new things. If you announce a new initiative, they will probably be the first to volunteer or to ask for more information. They, like the pioneers of old, get a thrill out of being the first ones to get to the frontiers.
- *"Yes" people.* These are people who "do as they are told." If the boss or the company asks them to do something, they feel obligated to do it. They seldom volunteer, but they almost never refuse.
- *Crowd followers.* These are people who want to be on the winning side and are not sure which side that is until they see one side clearly winning. When they see trends toward something new taking root and happening, they want to be part of it. Until they are sure it will work, they want to wait and see.
- *Skeptics.* Skeptics can be convinced only by their own careful, thorough (and usually slow) analysis of the new initiative. No amount of hype, marketing, momentum, or popularity will convince them. But, once convinced, they can become staunch advocates and valuable allies to the cause. Also, most good skeptics have a secret following or public following of people who admire their analytical skills and value their conclusions. Winning over a skeptic often means winning over their followers also.
- *CAVE people.* CAVE is an acronym for citizens against virtually everything. These are the people who will help you with your new initiative when hell freezes over and pigs fly. Not really! Even if you could arrange those two happenings you could not count on these folks. They do not join, do not help, and do not care if you succeed. They just want to get their check and go home.

The reason we suggest that you think about your people in these categories is to help you realize and carry out the most effective change strategy. Many people think, "If we could only get the CAVE people won over the rest would be a piece of cake." That is true except for the fact that you *cannot* get the CAVE people won over because they arc CAVE people! If you fall for that approach, what usually happens is that the organization sees a little group of change agents doing mortal battle with a little group of CAVE people, and they either decide that this is meaningless or take sides. Either way is disastrous for your initiative.

Effective change follows an intelligent strategy that starts at the right place, gets quick wins, and builds momentum toward the goals. The way to do this is to start with the pioneers who want to help anyway. The ideal first wave of change agents is ALL pioneers and NO CAVE people! When you run out of pioneers, go to the "yes" people and ask for help. They will say (did you guess it?) YES! When you have enlisted this group, you will find that the crowd followers are beginning to see the trend and will start to be willing to join and help. Somewhere in this time frame also the skeptics will begin to complete their painful analysis and decide that your initiative makes sense and deserves their help. As they begin to join, their followers will follow.

Now, it is EVERYONE against the CAVE people. Is it time for an all-out attack? NO! You do not win over CAVE people, and you do not do battle with those who have nothing to lose. Almost no initiative wins over 100% of the organizational population. The vast majority is enough for success, and you will begin to realize that even CAVE people are influenced by excellence. Not in the same way as everyone else, but indirectly, they will realize that they cannot be a part of an excellent organization unless their own performance becomes more excellent (or less visible). We have seen groups of CAVE people have their own secret safety meetings in which they say they do not want to join STEPS but realize that it would sure look bad if they were the only group getting injured.

Questions for STEP 1.9

1. Do we already have a group of change agents we can utilize: existing safety team or committee, safety champions or union safety reps, behavior-based safety (BBS) steering team and observers, or others?

2. Do they fit our needs or do we need to modify this group or form a subgroup?

3. If we have no suitable group of change agents, how should we best get one: read ahead and form our SET for STEPS, form a temporary team, assign a person from each work group, or other?

4. What training will our change agents need in addition to a thorough understanding of our safety strategy?

5. Do we need to develop SRRRs for them or can we define their roles more informally?

6. What publicity do we need to make sure that everyone knows who they are and what their function is?

7. What do managers and supervisors need to do to make sure that they are enabled to fulfill their roles?

STEP 1.10   MEASURE/ADJUST

Background Materials

Even the best of strategies are just plans and projections of what can and should happen in the future. The future is unpredictable and can possibly keep those strategies from working perfectly. Good strategies are precise, but flexible. They also include ways to measure their own progress and to adjust if the metrics suggest that they need change. In safety, we have a history of simply measuring results and trying to manage our efforts with lagging indicators. In achieving sustainable Safety Culture Excellence, we need to move beyond that view of measurement and learn to measure effort, progress and process, as well as results. The results are certainly what we want to impact, but the other metrics are useful in helping us to accomplish that goal. Think of the speedometer in a car and the tachometer. The speed is what we want, but if the engine is not turning the right RPMs and the transmission is not doing its job, we would not accomplish our goal.

Think of your safety strategy in three parts:

1. Are you working your plan?

2. Is your plan working?

3. Is working your plan producing the desired results?

1. *Are you working your plan?*   Your safety strategy will call for actions. They may include training, communication, forming new organizational structures, recruiting people to help, developing plans, holding meetings, and so on. A simple metric for these types of actions is simply calculating a percent complete. Make a list of assignments: who is going to do what and by when? You can divide it into sections or types of activities if that helps, such as the categories mentioned previously. Some of the activities may be ongoing, monthly, weekly, quarterly, and annually. Make a list and a grid and tick off items as they are completed. Ask for regular reports from those assigned and calculate what percent of what you targeted to get done actually is completed.

If possible, develop evaluations of the quality as well as the quantity of activities. These can be evaluations and do not have to be precise or discrete metrics. Remember the adage that "an imprecise measure of the right thing can be better than a precise measurement of the wrong thing." Always try to measure what is important, not just what is easy to measure. If you are asking individuals to make a personal value judgment of the quality of an activity, be sensitive to how people can influence each other. Let the evaluations be completed privately and discretely and not in group settings. You can report and share the evaluations in the group, but do not let the group influence the individual evaluations.

If your metrics indicate that any part or parts of your plan are not happening, adjust. These adjustments to your plan can be changes to the plan itself or simply changes in how you carry out the activities. You can change who is assigned, assign more people to help, or change the timeline for carrying out the activity. If the activity is completed, but evaluated to be of poor quality, you can assign a new person or team to take it to the next level or modify it. When you adjust your planned activities, be cautious not to create an atmosphere of blame or rating personal performance. Each person involved is likely a volunteer (and definitely a member of the culture you wish to improve), and their performance should be appreciated, not punished. Evaluations should be of strategic progress and not just personal contribution. Adjustments should aim at accomplishing the, as yet, unaccomplished strategy, and everyone should remain united, cooperative, and appreciative of all contributions toward success.

2. *Is your plan working?*   There is a set of metrics called key performance indicators (KPIs). These are measurements of the impact of your activities. If you have an activity to communicate, your KPI might be to measure how many people can remember or recite the information communicated. If you have formed a group and asked them to hold meetings, how many meetings have they held and what was the percentage of attendance? If you have held training, what percent of your target number of people have been through the training and what percent of them can demonstrate competence in what they were trained to do? If you are trying to change perceptions, have you administered perception surveys or held focus group meetings and, if so, how much have perceptions changed since the last measurement? If you have asked workers to focus on certain precautions, are they doing so more regularly? In short, are the activities in 1 ("Are you working your plan?") producing the desired changes?

If your KPIs indicate that your plan is not working or not working well, adjust! If your training is not achieving the desired capabilities in your workers, you can either adjust the training itself or change the persons delivering the training. If communication is not working or working well, you can change the message, the media, or the messengers. Adjustments in your plan may entail adjustments in how and who is working your plan. The overall decision is whether or not your plan is producing the desired changes or not; but do not overlook ways in which the plan could be made more effective or efficient even if it is working.

Remember also that even though there are many performance indicators, only a few of them are truly "key" indicators. Do not fall into the trap of thinking that a result that is upstream from your ultimate desired result is automatically a performance indicator. Performance is activity, and result is consequence. The KPIs are the activities most directly related to the desired results by cause-and-effect relationships. Small changes in true KPIs will produce visible changes in results as quickly as statistically significant results data are available. If you can change a process and not visibly or significantly impact the result, this indicator is not "key."

Case Study: We worked with an organization a few years ago that noticed that a high percentage of their accident reports pointed to inadequate training as a causal factor. They decided to try an experiment in which they increased the amount of training delivered to a targeted group of employees while keeping the training the same for another group as a control for the experiment. They found that the employees with increased training had an almost identical rate of accidents to the group that had no increase in training, and yet training was still cited as a causal factor on many of the accident reports. What they discovered was that the "quantity" of training made no significant difference in performance or in the results. Their next experiment attempted to increase the quality of the training. The better trained group virtually eliminated their accidents while the control group remained about the same. The organization determined that the quality of training (as measured by training evaluations) was a much better KPI than the quantity of training.

3. *Is working your plan producing the desired results?* The third part of measuring may already be accomplished in whole or in part. Your goal may be to impact your lagging metrics that you are already measuring. If so, make sure that you understand statistics and accurately interpret the impact and trends of your data. Misinterpreting data can lead to false conclusions about whether or not the plan is truly producing results. If the team or group looking at this data has no formal training or knowledge of statistics, consider adding an ad hoc member to the team who does know statistics and can help the team better understand and interpret their data.

The goal of multiple metrics is not simply to have more data. The goal is to develop a deeper understanding of what effort and which strategies produce the desired results. The real challenge of excellence is not just to reach it but also to understand and do the things that will duplicate the excellent results year after year. By measuring the accomplishment of the strategies or plans and the effort going into them, it becomes more and more clear how to impact the results in a positive way. This deep understanding is what W. Edwards Deming would have called "profound knowledge" of your culture and processes that comprise your safety strategy.

When excellent results are being achieved, it is easy to simply think that you are working your plan and that your plan is working. Further analysis can and often does reveal that some efforts are driving the results and others are not. It is important to constantly analyze and remove unnecessary or "nonvalue-added" steps. The late business guru, Peter Drucker, often reminded us, "Nothing is less productive than to make more efficient what should not be done at all." It is not enough to be successful at any cost. The goal should be to achieve success with an efficiency of effort and resources.

Questions for STEP 1.10

1. Are we working our plan?
   (a) Are following the right path?
   (b) Do we have the right people involved?

(c) Are our people doing what they need to do?

(d) Are there barriers to working the plan?

2. Is our plan working?

(a) Are we hitting our KPIs?

(b) Do people know what we are doing?

(c) Are we changing the way we do things?

3. Is working our plan producing the desired results?

(a) Have we improved the lagging indicators?

(b) Have we improved the culture?

STEP 1.11   CONTINUOUS IMPROVEMENT

Background Materials

True excellence is a fragile state. Unfortunately, it is not a fortress you can reach and rest safely inside. It is not a title you can attain and hold it forever without further effort. Excellence is a state of constant vigilance and continuous effort. The goal of the ongoing efforts should not be to maintain the status quo or to simply keep from backsliding. The goal needs to be continuously improving. When you reach your best, you must conceive a better goal and strive to achieve it. It is only by such efforts that excellence can continue to exist. Anything less than continuous improvement is less than excellent.

For this reason, the safety culture and the organization should continuously stretch the goals, expand the capabilities, and enlarge the amount of participation in safety. Every aspect of Safety Culture Excellence should be visited, addressed, and periodically revisited to explore ways to make it better. This is the journey you will begin as you complete your safety strategy and move on to the STEPS methodology. As you do so, remember not only to always improve but also to always take the improvement one step at a time. Trying to do everything or too much at once is a major cause of failure in safety-improvement efforts.

One of the famous generals of World War II was George Patton. He was often quoted telling his officers that he never wanted to hear anyone say that they were holding their ground. He believed that moving constantly ahead was the best and possibly the only way to win the war. He was a controversial figure, but he also accomplished some incredible victories and set some records that have never been equaled. In your battle to overcome accidental injuries, form a culture that is always advancing and never simply trying to hold their ground. Doing so is the direct path to excellence.

In the sections to follow, we will share models of what a good safety culture looks like and what it can do. We will tell of strategies that have worked for other organizations and some that did not. We will tell you some stories of these experiences and how we discovered some of the concepts and methodologies that make up STEPS. Please do not take this book as an absolute and only formula for success. You will discover other capabilities that can help make your safety culture more excellent and other methods to acquire or expand those capabilities. Excellence is a journey! Welcome fellow traveler!

Case Study: We led a major safety-improvement initiative for a petrochemical site. They called us because they were the worst safety-performing site in their division, and the management was being pressured to improve. They were not bad at safety. They benchmarked quite high in their industry, just not in their division of their company. They had recently increased the size of their safety department to try to address the issue and that had had some impact on improving results. The initiative we implemented increased involvement of the workforce and focused on some strategies that had not been tried. In less than a year, the site had passed the best performing site in the division in safety and had won the company award for best safety program and best safety improvement. The site manager called us into his office and thanked us for what we had done and announced that he had one parting question, "Why do I need any safety professionals, much less an expanded department to maintain what you have started? After all, it is mostly employee involvement!" What do you think would have happened if we had advised this manager to dismantle his safety department and try to hold his ground?

Questions for STEP 1.11

1. What is the trend in our safety lagging indicators: improving, staying the same, or getting worse?

2. What parts of our strategy most directly impact our lagging indicators?

3. How could we further improve one or two of those strategies?

4. How can we track this effort and see if it is having an impact on our lagging indicators?

Milestone 1 Application: When your leaders have finished these two workshops, they will have sets of information and insight from which they can formulate a safety strategy. The strategy will include the 11 elements in this workshop. In addition to a strategy, your leaders should have a much better grasp of what organizational safety is and what it takes to improve it. But the journey is not finished, and it is not time yet to roll out the new strategy. The assessment that is the next milestone on the journey will provide additional information that will impact how the strategy is applied. Again, it is important that the safety strategy be where the organization wants to go in safety and not just a set of Band-Aids to cover the wounds found in the assessment.

ASSESSMENT

Status quo: Latin for the mess we're in.
—*Jeve Moorman*

The assessment of existing safety culture should not be viewed either as a verification of the safety strategy or as information to modify or improve it. Measurement is a tool for understanding, and an assessment is a kind of measurement. Your organization is considering a journey of improvement. The first step of navigation is determining your current position. A good assessment tells you your starting point and allows you to measure progress as you travel.

You should carefully consider who should conduct the assessment activities. In some cultures, outsiders are viewed with distrust, and information can be withheld. In other cultures, there is an intimidation factor from people within the organization as they have the power to impact worker's position, pay, and employment. The gathering of accurate and complete information not only involves techniques but also uses the right person or persons. If you contract an outsider to do your assessment, make sure that the methods and instruments are customized to your needs and that the consultants are professional and experienced. If you choose to utilize organizational personnel, make sure to choose people who are trusted, good communicators, and who will not get defensive if workers criticize people or processes.

The assessment consists of four activities: an evaluation of existing safety programs, administering a perception survey, holding interviews with a representative cross section of employees, and performing a Pareto analysis of safety data. These four tasks can be divided among individuals or all performed by the same person or persons.

Goals: To understand and appreciate the current status of your safety culture

To determine what is currently influencing the culture

To evaluate existing safety teams or committees for possible use in STEPS

STEPS to Safety Culture Excellence^{SM}, First Edition. Terry L. Mathis and Shawn M. Galloway.
© 2013 John Wiley & Sons, Inc. Published 2013 by John Wiley & Sons, Inc.

To evaluate existing safety programs for possible improvements

To establish a baseline for measuring progress

Methods: Evaluation of existing safety programs

Perception survey

Interviews with individuals and focus groups

Pareto analysis of safety data

STEPS: 2.1 Evaluation of Existing Safety Initiatives

2.2 Perceptions

2.3 Interviews

2.4 Safety Data Analysis

STEP 2.1   EVALUATION OF EXISTING SAFETY INITIATIVES

Your organization is already working on safety. You very likely have a safety professional or designated safety leader, a safety committee or team (or more than one), safety training, safety meetings, and perhaps other initiatives such as Voluntary Protection Program (VPP) or behavior-based safety (BBS). Performing an evaluation of these initiatives can help you better understand the current status of your safety performance and your existing safety culture. Such an understanding can help you to better determine your improvement opportunities and to provide a baseline to measure your progress.

Part of your evaluation of these initiatives can come from your perception survey and interviews, but it will help to actually visit and talk to people involved in some of the initiatives and activities. A firsthand knowledge of the programs can also help you customize your perception survey to give you more focused and complete information. (We will recommend how you customize your perception survey in the next section.) Plan to sit in on committee meetings for each initiative and/or the site safety committee meeting. Talk to the members of the committees and ask them their opinion of the initiative's effectiveness and impact on safety. Attend a cross section of safety training sessions and evaluate the training program. Talk to people attending and get their opinions of how effective and useful the training is. Attend safety meetings and evaluate how they are presented and how they are received. Look for the level of interaction and participation and talk to people about their opinions of them.

When talking to committee members, consider asking questions like these:

- Did you volunteer to be on this committee or were you selected some other way?
- How long have you been a member?
- What do you think is the main goal of this committee?
- How well do you think you are achieving that goal?

- Do the members of the committee get along and work well together?
- What do your fellow workers think about this committee and how it helps safety?
- What is your biggest challenge or opportunity?
- What do you think are some of your successes so far?
- How do you measure the effectiveness of your initiative?

Case Study: We visited the committee meeting for a BBS initiative at a client site and questioned the members about their process. We asked them the last question mentioned previously (How do you measure the effectiveness of your program?), and they responded that they had excellent attendance at their meetings, that they were regularly hitting their target number of observations, that 100% of their observers were active, and that they had a monthly meeting with management to give them a progress update. We asked them a follow-up question, "How is your process impacting accidents at the site?" They had no idea. We too often find that safety programs led by individuals or teams can become what we call "process oriented" rather than "results oriented." They had forgotten that the whole purpose of their program was to reduce accidents.

When talking to people at training sessions, consider asking questions like these:

- Have you taken this exact same training before?
- Does this information really help you do your job safer?
- Is there too much information to remember once you get back to work?
- How could we make this training better?

When talking to people at safety meetings, consider asking questions like these:

- Is this meeting a good example of what they are usually like?
- Do these meetings help you do your job safer?
- Do you feel free to ask questions in these meetings?
- How could we make these meetings more useful?

Feel free to add or otherwise customize these questions to your organization and probe with follow-up questions when you discover interesting or valuable information. Take notes and tally the responses you get.

STEP 2.2   PERCEPTIONS

An important aspect of your starting position is what people think about your current safety culture, efforts, and results. What a person thinks is not necessarily accurate, and members of a group may differ in their perceptions of the same issue depending on their own personal experiences and other factors. However, as your STEPS process begins to take effect, the progress will be reflected in changed perceptions across the population as a whole. Taking a beginning sample of perceptions will allow you to track these changes as your process progresses. It may also help you identify issues on which to focus future action plans where they will make the most difference.

Perceptions can be measured in a number of different ways. You can administer a survey via computer or pencil and paper, ask questions in interviews and tally results, or use various computerized equipment to capture responses to questions delivered either verbally or via written media. Formal perception surveys can be purchased or custom developed. We recommend that you do not attempt to use purchased surveys not designed specifically for the STEPS process. Such surveys will include perceptions irreverent to STEPS and omit important perceptions. They will often use terminology other than that used in STEPS to convey the same concepts, which can be confusing to workers and skew the results. Such off-the-shelf surveys also tend to be expensive and benchmarked with groups that do not reflect STEPS or even safety-culture improvement efforts. For these reasons, we recommend that you either custom design your own perception survey or purchase one designed specifically for STEPS.

It is important to remember that a measurement of perceptions is only one aspect of your starting place as a safety culture. Since perceptions can either be accurate or inaccurate, you can do what we call "grounding" your perception measurement by comparing it with reality. There are several ways to ground a perception survey:

1. Through interviews, determine if people accurately understood what the survey was asking and responded accordingly. In a recent meeting with a potential client, we were told that a certain manager got low ratings for his communication skills because his direct reports rated communication lower than did people in other departments. When interviewed, the employees responded that their manager was a great communicator, but they felt that other managers were weak communicators. The question basically asked how well managers in the organization communicated with their direct reports. They responded based on their overall organizational perception, but their responses were misinterpreted to indicate poor communication from their manager.

2. If you measure perceptions of risks, you can compare and contrast perceptions with data from other sources. For instance, if you ask workers what they perceive to be the most likely type of accident that could happen to them or the part of the body most likely to get injured, you can compare their responses to accident-investigation data to determine the level of accuracy. Remember that risks can be controlled by focus and that the greatest risk may be already better controlled than the risk most often resulting in accidental injury.

3. If you measure perceptions of the effectiveness of programs or processes, you can also compare these with data. For example, if you ask about the speed with which maintenance responds to safety-related work orders, you can compare actual response times to these perceptions.

4. If you measure perceptions about the effectiveness of safety training, you can compare this perception with the number and percentage of accident reports

that list "lack of training" or "inadequate training" as a contributing factor in the accident.

If you ground your perception measurement, you can determine a percentage of accuracy. As you work on the STEPS processes, such as communication, you can track the accuracy of perceptions increasing as well as the changes in perceptions. As your process progresses, people's perceptions will change about effectiveness and efficiency of safety efforts, performance of managers and supervisors, and the capabilities of workers to collaboratively drive improvement.

Case Study: One of our clients who hired us to help them better assess their safety culture told us of an earlier experience in which they had purchased an off-the-shelf perception survey and given it to their workers. The workers in one department had a very bad perception of the usefulness of their safety training that was being delivered by a new member of their training department. The rest of the departments perceived their training to be significantly better. There was an immediate response that the new trainer must not be doing his job, and a suggestion was made to transfer or terminate him. Cooler heads ordered an investigation of the situation. They found that the new trainer was delivering twice as much safety training to this department as the other trainers were to their departments, that the department had not had a recordable injury since the training program matured, and that workers had heard the same message until it was sounding redundant and boring. However, they had also internalized the training, started regularly taking some precautions, and virtually eliminated their accidents. The perception survey results did not tell the whole story and almost caused an inappropriate reaction. Ungrounded perceptions can be accurate, inaccurate, or incomplete, and you cannot determine which simply from the survey data.

Guidelines for Administering a Perception Survey: If you use a customized or purchased perception survey, it is important to observe some guidelines for administering it to the population of your organization:

- Keep surveys anonymous by not including employee names. Be sure that other variables do not triangulate back to the worker. For example, if you have no name but include shift, time, work station, and employee position, then anyone can determine the employee by the variables without needing the name.

- Time the administering of the surveys to avoid other issues that might impact perceptions. For example, do not conduct a survey too soon after announcing a layoff or reorganization. Avoid having workers stay over after a long shift or come in on days off to complete them.

- Make sure that you include enough people to constitute a statistically significant sample of your total population. If you are not sure how to do this, seek help from someone with an education in statistics. If you do not have an engineer or trained professional at your site and do not use a consultant, seek the statistics advisor from a nearby college or university. An alternative is to give the survey to everyone. ("Everyone" is always a statistically significant sample.)

- Make sure that you include people from every department, shift, task, tenure, and other groups in the ratio of their total percentage of the whole to make the sample representative of the total population. Again, an alternative is to give the survey to everyone.

- Eliminate any intimidation or influencing of answers by making sure that bosses or fellow workers are not watching or prompting others as they complete the survey. If you use computers or pencil-and-paper surveys, you can have people complete the surveys in groups as long as the groups are monitored to avoid problems.

- Collect paper surveys and secure computerized surveys so that none are lost or seen by anyone other than the participant and the data manager.

One of the reasons for measuring perceptions is to be able to manage them in the future. Unmanaged perceptions of safety create lack of focus and potentially conflicting efforts. We are not talking about manipulating perceptions, but correcting them. We do not want people to think our way; we want them to think correctly about risks and precautions. In the absence of good safety communication, workers tend to rely solely on their own personal experiences and the experience of those closest to them or those most visible. Such perceptions will always be an incomplete part of the whole picture of safety. These perceptions are also almost always based on statistically insignificant samples of data. Such perceptions tend to cause workers to put their safety efforts into less impactful precautions. Maintaining the proper focus can be a powerful tool in safety excellence and keep efforts producing the maximum results.

Caution: Measuring perceptions can be challenging. When designing and administering your perception surveys, please consider the following challenges and make sure that you do not diminish your own efforts or allow them to fall prey to common problems:

1. *Select the right perceptions to measure.* Most off-the-shelf perception surveys measure standard perceptions that are not customized to the site using them. If you are focusing on certain issues, measure how workers perceive those issues and progress toward those goals. Avoid the standard "How are we doing?" kind of questions and really drill down into what matters and what is current. We continue to recommend a "do-it-yourself" approach to perception surveys for most organizations to allow for this kind of meaningful customization.

2. *Realize that perceptions are volatile.* There are a lot of things that can impact perceptions in the short term. Avoid measuring perceptions right before or after major changes or announcements. Do not interrupt lunches or breaks or ask people to stay after work to impose a survey. Try to measure perceptions during normal times and avoid anything that would cause temporary interruptions, anxiety, or fear.

3. *Use the right terminology.* Most off-the-shelf surveys use standard terminology that might not fit your organization. It is confusing when the survey refers to "supervisors" when you call them "foremen." The standard terms often

misunderstood include references to organizational positions, meetings, training, or specific programs or initiatives. Make sure that you are using the terms that are in common use among the troops.

4. *Be careful how you phrase things.* Simple wording is more often effective in surveys than complicated phrases. Many surveys include such wording as, "Safety is an integral part of my job"; or "Our leaders regularly display their support for safety." Such phrases are not terribly complex but can be interpreted more than one way or simply misunderstood. Stick with simple and straight wording such as, "My boss expects me to do work safely"; and "My manager talks about safety in every meeting."

5. *Avoid intimidation and maintain confidentiality.* Make sure that supervisors do not stand over workers while they fill out surveys and that it is not too easy to trace a particular survey back to a particular worker. Administer the surveys in private and collect them anonymously.

6. *Avoid "haloing".* Haloing is when a worker expects he should basically agree with everything and can mark the approximate same response to every question or statement. Reverse the order of some where the expected answer is to disagree and make people think about each answer as much as possible.

7. *Act on previous surveys before administering new ones.* Failing to respond to a survey or failing to communicate how you have responded to a survey is a good way to build resistance to the next survey. If you are not going to use the data, why should workers give you more?

8. *Avoid misinterpretation.* It is easy to look at responses on a survey and jump to conclusions. Make sure that you are objective and not assuming when you interpret survey responses. One of the best ways to ensure this is to interview respondents and ask probing questions. Do not assume just because workers rated a supervisor lower on safety than another that it is definitely because of some personal fault or failure.

9. *Measure with the right frequency.* Because of the extreme expense of most off-the-shelf surveys, most organizations do not use them frequently enough to really understand how they are impacting perceptions. Shorter, more customized surveys can be utilized more frequently. The impact of programs or initiatives can be more accurately measured with frequent surveys. Many organizations perform short surveys quarterly and even more often for special needs.

10. *Be sensitive to language and literacy issues.* Provide the survey in all languages spoken and, if literacy is an issue, do the survey together verbally while each participant marks their responses to avoid embarrassment while still getting accurate data from workers with literacy issues.

11. *Trend and/or benchmark accurately.* Make sure that you use good statistics when defining trends or benchmarks from survey results. A few data points do not make a trend, and a one-time survey is questionable to compare with another site or company with great accuracy. Take the results as tentative until you have enough data to really see how it is trending.

If you decide to design your own survey, we recommend the following format and scoring. We have tried many different ways and hired many different experts, and this is the one we keep coming back to. Here are the design issues we think are most important:

1. Measure perceptions about each of the elements of cultural chemistry and climate mentioned in Milestones 4 and 5. For example, the first element of safety chemistry is "passion." You might want to measure the perception of how strongly workers feel or care about safety and each other or how much they think others care. Each element should have more than one aspect to be measured.

2. Make statements rather than questions. For example, do not ask, "Do you feel a strong commitment to keep your fellow workers safe?" Make a statement like, "I feel a strong commitment to keeping my fellow workers safe."

3. Ask people to agree or disagree with the statement. Give them a range of agreement and an out if they do not have a perception about this topic or do not have enough information about it. We prefer the following range of responses: strongly agree, agree, ?, disagree, and strongly disagree.

4. Give these responses a numeric value but do not use a Likert scale where the number is only a label for one of these responses. If the desired response is to agree, then give "strongly agree" a value of +2, "agree" a value of +1, "?" a value of 0, "disagree" a value of −1, and "strongly disagree" a value of −2. If the desired response is to disagree, reverse the values. This way you can give responses a value and trend them over time. An ideal response will be +2, a good response will be +1, the percent uncertain will be the percent that "?" was selected, and undesired responses will be negative numbers. When you administer a perception survey to a number of people, you can average their responses to each statement and category of statements (elements of chemistry and climate) with a scale that ranges from 2 to −2 and calculate a percent of uncertain responses. As you progress toward Safety Culture Excellence, you can see perceptions that need to improve responding or not to your initiatives.

5. Have clear instructions on the surveys about how to fill them out in case they are simply handed out instead of administered properly. If you administer them via one of the many computer programs designed to do so, the instructions will already be there along with a tutorial in many cases.

6. Keep them confidential; no names of those filling them out. We know that many of you would like that data, but take our word for the fact that taking names will skew your data more often than not and will sometimes completely destroy your ability to measure perceptions. Remember that measurement should help you understand, not to place blame or overreact. You are looking for cultural issues that you can improve and the overall progress of the culture, not the rotten apples in the barrel.

7. Do not make the surveys too long. Some organizations even measure only certain elements that they are focusing on improving. Short surveys administered more often usually produce better improvement than longer surveys administered yearly or less often. Generally, the longer the survey, the less often organizations can effectively administer them.

8. If this seems overwhelming or out of the scope of your organizations expertise or normal operations, there are consultants to help you as well as premade surveys for STEPS available from ProAct Safety®*.

9. Compile overall survey responses and share the data with those who completed surveys and with the entire organization. Publicize any specific responses to the survey along with results produced and thank people for their participation. Failing to share data and responses can create a barrier to future survey participation.

10. Chart the progress of your surveys over time and use the data to measure progress and to identify issues that are not perceived to be improving. Always make sure to ground the survey results as described earlier before responding. Action plans addressing problems that are perceived but not real may not be very effective.

STEP 2.3   INTERVIEWS

The data obtained from interviews are often richer and more complete than data from perception surveys alone. It can be equally more difficult to compile and use depending on how you measure and record it. If you simply talk to people and ask their opinions and observations, you will get a "feel" for what the culture is like. You can generalize and try to describe what you sense from what you have heard, but such description is almost always viewed as very subjective and nonspecific. Recommendations from such data are harder to implement and to convince key players to support.

In contrast to the informal discussions you conducted to evaluate your safety programs, these interviews should be more formal and timed with your perception surveys. Your initial assessment should become a template for future mini-assessments to measure your progress.

There are techniques for conducting interviews and recording interview data that will help to avoid the problems described previously. The more uniform you can make the interviewing process without missing or suppressing the flow of information, the better. Some interviewing guidelines include the following:

1. Schedule interview groups of exclusively the same level of employee: middle manager, superintendent, operator, maintenance, and so on. Do not ever have anyone's direct "boss" in the same session. Even if you do not think doing so would intimidate those involved, do not take the chance.

---

* ProAct Safety is a registered service mark, U.S. Registration No. 4,011,455.

2. Schedule groups of a size that will allow for multiple points of view but are not so large that everyone will not be able to give their point of view.

3. Before starting the interview, set attendees at ease by stating that names are not being taken, and no one will be quoted. The information will be used to help design a safety-improvement initiative and that everyone will have future opportunities to offer input. Also remind attendees that just because someone offered an opinion does not mean that others must agree. Encourage opposing points of view and stay on the same question until you have gotten all the input from the group.

4. Have a predetermined set of questions and ask them in the same order in each interview session. Make them parallel to the perception survey categories so you can easily compare the interview data to the survey results. Keep track of how many attendees concur on answers. Use only one question list for each level interviewed. This will keep you from having to tally answers from different sheets.

5. As interviewees answer your questions, tally their answers with tick marks. If you are asking a "yes or no" question, you will get a tally of yeses and noes. If it is a more open-ended question, leave space to take notes on the various responses, and when you get duplicate responses, simply put a tick mark next to the original.

6. If you discover an issue that you deem important that is not on your list of questions, add it and ask future groups. Probe deeper into issues where you feel there is not enough information coming out from the planned questions.

7. Some of our favorite, nonstandard questions for interview sessions:

   (a) What is the most likely kind of accident that could happen to you in your job? (Not the most severe, but the most likely.)

   (b) What one thing, if everyone here did it differently, would make the biggest difference in safety?

   (c) Can you obey all the rules, follow all the procedures, wear all your personal protective equipment (PPE), and still get hurt?

   (d) How big a risk would you have to take before one of your fellow workers would say something?

Case Study: In one of my earliest assessments, I ended an interview session with the impression that the group of workers interviewed was extremely negative and critical of the site's safety efforts. I had administered the perception survey at the beginning of the interview and pulled the forms out and laid them side by side on the table. There were 12 people in the session, and two sets of responses on the survey were negative. The rest of the responses were positive and about average to the site's overall response so far. I had almost let two outspoken and negative people become the voice of the interview session. Ever since, I make sure to administer the perception survey in the interview sessions and compare the responses after each interview as a reality check. By the way, when you do the perception survey in your interview sessions, you have complete control of the environment in which the surveys are administered. You can assure that there is no intimidation, rushing, or loss of confidentiality.

STEP 2.4   SAFETY DATA ANALYSIS

Sites with mature traditional safety programs and processes may already have much of the information needed to satisfy this step. Even though we will introduce some new metrics to measure safety culture elements and progress, the traditional measures of safety (accident rates, severity rates, accident investigation information, etc.) will continue to be relevant, especially if they are data rich. Many good accident-investigation processes already tally the accidents and near misses by accident type and part of the body injured. Such data are needed in this analysis. These data can be used to ground the perception data and see if workers accurately perceive what their greatest risks are. It can also be used to see if safety efforts are focused on the areas of improvement with the most potential. If such data are not requested on the accident-investigation form, it should be added. If it is on the form but not tallied or reported, it will need some work to total the categories for a comparative analysis.

In addition to the more traditional analysis of accident and near-miss data, we recommend that you utilize another type of Pareto analysis that is designed to discover which precautions and conditional changes could have the most impact on accidents historically. The form and instructions for this analysis are contained in Appendix B. This analysis can also be utilized to check the accuracy of perceptions, but it plays several more critical roles as well. This analysis will be the tool that ultimately helps the organization to reach Milestone 6, Capability. Once you have taken the steps to determine your current status and established the kind of climate and chemistry necessary to grow an excellent safety culture, the ongoing work of excellence will involve continuous analysis to enable targeting and prioritizing safety-improvement opportunities. This form of analysis is the enabler for that process.

Milestone 2 Application: Combine the information about your existing safety programs, perceptions, interviews, and safety data analysis to get a good picture of the starting point on your journey to Safety Culture Excellence. This is the baseline for measuring your progress and a way to identify areas that are not improving at the same pace as others. Progress through the remaining STEPS in order before responding to issues identified here unless they are critical. You must take the journey a step at a time and take foundational steps that will build a base of support for your safety efforts before moving on. If you do not, you run the risk that your efforts will be undone by your own lack of support and structure. The step you have just taken to better understand your existing safety culture will help you as you begin the next step of developing clarity in your safety efforts. The assessment should be repeated yearly to measure progress, adjust efforts to changing issues, and motivate the process.

If your organization already has a robust safety auditing process, you might see if it accomplishes the goals of this milestone. If it does not, you might consider modifying it so it does. Many of our client organizations had safety audits that could be moderately altered and supplemented to do everything needed for this assessment. However, most of our larger clients only audit a site every other year, which is, in most cases, too infrequent to maintain the momentum toward Safety Culture Excellence. In several cases, they utilize the regular audits and supplement them with mini-audits on the off years, and that works well for them.

MILESTONE *3*

CLARITY

If you can't explain it to a six-year-old, you don't understand it yourself.
—Albert Einstein

After completing Milestones 1 and 2, the leaders of your organization should have a safety strategy and a very good idea of the current status of the safety culture. Now it is time to get the culture moving toward excellence. In the STEPS to reach this milestone, you will share the rationale for safety excellence with the general population of your organization, select or establish a team to steer the STEPS efforts, begin to align thinking about what safety is and how to improve it, and complete and implement the marketing campaign you started in your safety strategy.

One issue we find too often hampering safety is a lack of clarity. We tend to assume that everyone knows what safety is and how to make it happen. What we find in our assessments is exactly the opposite. Everyone has a different idea about safety, and most of them do not really know how to make it happen. Breaking safety down to something you can explain to a 6-year-old is not condescending; it is the way to create deep understanding and profound alignment of efforts. When everyone is on the same page thinking the same way and taking the same step for the same reasons, excellence begins to happen.

Goals: To designate or establish a Safety Excellence Team (SET) to steer the STEPS process

To set clear expectations about the what and how and why of STEPS

To align thinking about safety (get everyone on the same page)

To define crucial terminology and methodology

To begin to market the safety-excellence journey

Methods: Evaluation of existing safety committee and/or formulation of a new one

Training

Workshop

Workforce briefing

STEPS to Safety Culture Excellence[SM], First Edition. Terry L. Mathis and Shawn M. Galloway.
© 2013 John Wiley & Sons, Inc. Published 2013 by John Wiley & Sons, Inc.

STEPS: 3.1 SET Structure

3.2 SET Strategy Briefing

3.3 SET Clarity Workshop

3.4 STEPS Employee Briefing(s)

STEP 3.1 SET STRUCTURE

In order for STEPS to be successful, it needs to involve the right people in the right structure. Many forms of government give structure to the national culture through forming government entities that represent the people from different regions of the country. When you begin STEPS, you may also need a mini version of a parliament or congress. Most organizations already have safety committees or process steering teams for safety. Such a team or committee can possibly be utilized for STEPS, but we suggest that you analyze the makeup of the team to ensure that it meets the criteria for success. We strongly encourage using existing structure whenever possible and avoid creating new teams or committees. This may involve modifying an existing team or using a subgroup from a team to lead the STEPS journey. For simplicity, we are going to refer to this team through the rest of the book as the SET. You are under no obligation to call your team by this name, but you are free to use this title if you like it and if it fits your culture.

The site or organizational safety strategy should have been developed by the leaders. The continuous improvement and addressing of issues should be led by the SET unless there are compelling reasons not to do so. In some organizations, the safety issues involve proprietary company secrets and other forms of sensitive information. In others, the safety issues are not secret but involve complex science and technology that cannot be addressed by personnel without special training. In such instances, the safety efforts might rightly be led or supported by leadership or technical specialists who, in turn, involve others through a lower level SET. In most other cases, it is wise to involve members of the culture in the SET whose job is helping to reach Safety Culture Excellence.

What this means is that you need to have the right mix of people on the team. If the SET is to successfully lead STEPS, it needs to represent the rest of the members of the safety culture in position (manager, supervisor, team leader, worker, etc.), job knowledge, tenure, skills, and several other attributes. A complete list of the qualities needed in such a team can be found at the end of this STEP section. Remember that not every member of the team needs to have every attribute listed, but rather the whole team needs to have the range of attributes. To help you analyze your existing team or a hypothetical new team, there is an evaluation worksheet outlining the list of attributes that will allow you to list each team member and to rate him or her on each of the attributes to determine if you are lacking anything that might be critical to the success of the SET. This worksheet for team evaluation can be found in Appendix C.

The SET should ultimately have a rotating membership, but the term or turnover needs to be carefully managed. These team members will receive special training and acquire skills necessary to make STEPS successful. Losing too many and replacing them with untrained individuals can stall the process. However, letting STEPS become the exclusive property of a small and static team will alienate it from the very people it needs to represent and impact. Teams that rotate one member every 6 months to 1 year often have success in both maintaining a critical mass of needed skills and knowledge and, at the same time, allowing others from the culture to have a turn leading and learning the process. It is often necessary to have at least some of the original team members serve longer terms than normal to get the process started and to begin the cycle of rotation.

The principle of rotation can have exceptions. We have client sites where the SET members are relatively permanent. Most of these are smaller sites that have obvious safety champions and leaders from the work teams. However, some larger sites have legitimate reasons to also have more static team membership. These can include lean staffing, difficulties in crossing over shifts, union issues, and simply preference by the culture for stability in safety leadership. Test the waters with your culture and see what level of rotation, if any, is desirable and remain flexible if you need to change. The best indicator of the need for rotation is a larger number of volunteers to be on the team than you can use at once.

If a site manager is on the team, he or she can be the leader of the team as a natural order of organizational design; but some teams find advantages in rotating this responsibility also as the process matures and other team members grow in ability and understanding. Also, some sites have a frequent turnover of site managers and that can destabilize the SET if it becomes too dependent on the site manager and then has a change. Teams should involve workers from the shop floor or field to represent the workers from their areas or departments. Sometimes these worker members of the team can become an independent subteam to address specific work-place safety issues later in the STEPS process.

Within these teams and/or subteams, we recommend that you avoid, if possible, the traditional, committee thinking on leadership. You do not necessarily need a chairperson, cochairperson, scribe, and so on. If the entire SET or the subteam can be self-directed and rotate leadership responsibilities, that is usually best. If the members are determined to have the more traditional leadership structure, encourage them either to wait in selecting the leaders or to select the first leaders for only a short tenure of service. Often, the members who are not the best leaders are the first to volunteer and, if serving too long, can impact the effectiveness of the team.

If the SET or subteam has no natural leaders and struggles to do its job effectively, a facilitator can be provided. A facilitator is not a voting member of the team, but an ad hoc member who keeps the team on course and helps them to outgrow the need for facilitation over time.

Extremely small sites may use a single person to head up STEPS, but most sites will find it best to use a SET. We recommend that the SET size be no fewer than three and no more than 12. The main considerations in team size are (1) representation, (2) critical mass, (3) the ability to meet, and (4) the ability to collaborate without forming factions.

1. *Representation.* Since the SET will be addressing safety issues, it is crucial to have someone on the team who is familiar with the operational issues of

each area or department of the site. Another goal is to make every employee feel that someone on the team understands and represents their interests and specific issues. Some employees may have a working knowledge of more than one area or department and that can be used to minimize team size and still accomplish representation. The three considerations here are representation (everyone feeling like they have a champion on the team), expertise (someone on the team who has expertise on the work and safety issues of every major part of the organization), and involvement (the team members help create the process). As W. E. Deming so often reminded us, people support what they help create.

2. *Critical mass.* The SET should have enough members to survive a loss without the need for more training. A team of two who loses a member is almost unable to continue. A team of five who loses three members may also struggle. The SET should have enough members to survive turnover and be able to bring new members up to speed without outside assistance, if at all possible. A site with high turnover rates should consider this when sizing their teams or when selecting employees with more potentially stable jobs for the team. If turnover is inevitable, many sites choose to train alternate SET members to fill upcoming vacancies without the need for retraining.

3. *Ability to meet.* A team that cannot meet will never really be a team. When you think of representation, you must also consider the ability to meet. It might be an ideal representation to have a SET member from each department on each shift, but if they cannot meet without many members coming in on days off and others staying over from a previous shift, they might not be able to meet, or they might not continue to make such sacrifices to be on the team. Selecting the SET members is a balancing act between representation and ability to meet; but ability to meet must be the primary consideration. In larger sites, we have formed different SETs for different plants, departments, and even shifts. If you have multiple SETs, you should consider a structure where they can correlate and share ideas. Many large sites allow the leader of each SET to meet with the other SET leaders quarterly or semiannually.

4. *Ability to collaborate without forming factions.* SETs need to be able to work collaboratively, and that becomes more difficult with larger numbers of members. It is important to weigh this issue with the critical mass consideration and to find a solution that addresses both. If employees have had training in collaboration, team building, problem solving, and so on, this can facilitate their ability to work together more effectively in larger groups. However, a larger team is not necessarily a better team.

Historically, it is difficult for a team of 15 not to form 8–7, 9–5, or 10–5 factions. It is possible for a team of 12 not to form factions, but it happens far less often. As teams become larger, it can also impact their ability to meet without shutting down operations. These are two of many reasons we recommend that SETs not exceed 12 members.

Again, if a site is large and/or multifaceted (such as a multiple-product manufacturing site or sites with large maintenance or contractor groups), consider having multiple SETs for each of the various large factions. The more different each group's safety challenges are, the more practical multiple SETs will be.

Remember, the first consideration for team membership is to select a representative from each major area, department, or shift at the site, that is, production, maintenance, shipping, and so on. The SET will make decisions for the rest of the workers, so it is important for each major area to have a representative on the team.

Here are some real examples of SET makeup from different organizations:

- SET for a small manufacturing plant with a total population of 88 (six members): plant manager, plant production manager, plant maintenance superintendent, production worker, maintenance worker, and shipping and receiving worker.

- SET for a large paper mill and converting plant with a total population of 1241 (12 members): site safety manager, paper mill superintendent, converting superintendent, contract maintenance site manager, three paper mill workers, three converting workers, one contract maintenance worker, and one logistics worker (warehouse).

- SET for a medium services provider with a total population of 395 (eight members): site manager, site sales manager, site service manager, office manager, one outside sales person, one inside sales person, two service providers.

- SETs for a large chemical plant with a total population of 1805 plus an average of 350 contractors—SET for each of four shifts: shift supervisor, shift maintenance foreman, two workers; shift supervisors from each shift meets quarterly with site manager and site safety officer. Once yearly, all shift SET members attend this meeting.

Characteristics of Effective SET Members

Use the following guidelines as you consider employees as potential SET members. Remember that it is the team that needs these characteristics and not each and every member of the team. Members of the team can have talents in various areas and not in others as long as the whole team has enough of the desired characteristics to be able to function. As you consider the members for the team, use the chart in Appendix C to help you evaluate the overall team characteristics.

• *Willing participant*	Good team members cannot be coerced into participation. It is not necessary that each team member volunteer, but he or she must be willing to participate when invited. However, many good team members have expressed a level of skepticism at first. Being skeptical is common and should not automatically eliminate a worker from consideration.

• *Respected worker*	Team members must be taken seriously by both managers and peers. Be careful not to mistake popularity for respect.
• *Experienced*	Novices seldom have adequate knowledge of the site, the people, or the safety issues to be effective team members. Do not necessarily select all "old timers," but avoid anyone who has not been on the job long enough to really know the ropes.
• *Good safety role model*	Team members do not have to be accident free, but they need to be serious about safety. Often, employees who have been injured develop serious attitudes about safety and have great credibility when they speak about safety to other employees.
• *Open-minded*	Team members need to be able to learn and manage new ways of doing things. Some people are overly opposed to anything new or different. Such people can seriously hold back progress if they are put in a position of leadership.
• *Collaborative*	Team members need to be good team players. They need to work together and be able to come to consensus. Workers with reputations for stubbornness and not getting along with fellow workers should be eliminated from consideration.
• *Takes initiative*	Good team members should be self-starters. They should have demonstrated initiative on work projects or extra assignments.
• *Good communicator*	Team members should be able to effectively communicate with other workers and managers. They do not need to be professional speakers, but they should be credible and able to effectively express ideas.
• *Good problem solver*	Team members will be given data and expected to identify and solve safety problems. Workers with demonstrated ability to solve problems should be given strong consideration if they meet other criteria.
• *Personal organization*	Team members will need to manage multiple roles and priorities. Workers with the ability to get things done in an organized way often make good team members.

Other Change Agents

In addition to the SET, some sites designate others to assist with the STEPS process. These could be safety representatives in each work team, union safety reps or leaders, observers from a behavior-based safety (BBS) process, lead men on work teams, or almost any others that fit your needs. None are necessary but might be desirable, and they have helped some organizations progress faster.

Recruiting SET Members and Other Change Agents   If you are forming a new SET rather than utilizing an existing safety committee or modified committee, you can proceed two ways:

1. Ask for volunteers.
2. Select "voluntolds": qualified candidates that they either asked or expected to participate.

If you ask for volunteers, be sure to set realistic expectations. You may get more volunteers than you can use; in which case, some may have to be refused or asked to wait to rotate on to the SET. If you do not get volunteers from certain areas, you might have to resort to "voluntolds" to get the proper representation. Make sure that this is clear as you ask for recruits. Also, publish the list of qualities you are looking for in SET members and any other criteria for membership.

If you are selecting rather than recruiting, be sure to clearly define the duties and time requirements and ask for a commitment to participate as needed. Give a time of service range if possible. Even if you are using an existing committee, it is a good idea to describe the new requirements and ask for a recommitment from members to the new task at hand.

STEP 3.2   SET STRATEGY BRIEFING

Now that you have your SET in place, it is time to get them started moving through the STEPS to Safety Culture Excellence. You have already taken some STEPS without them, and they need to be briefed on what those were. You are enlisting them to help you and they will have input into the process in the future, but you might want to consider their input on what you have already developed. This is optional but has been successfully used by several organizations.

To give the SET input to your safety strategy, you need to outline what you have already developed and ask if they have additional ideas. It is crucially important to set the right guidelines and expectations for this exercise. Make sure that there is no misunderstanding that the leaders are looking for additional ideas and not promising to accept all or any of them. In fact, you should not decide to do anything with their input but take it under advisement during this workshop. The best way to approach this is to present the strategy and simply ask for comments or suggestions, but keep the session a presentation and not an editing session.

As you recall, there were 11 parts to your safety strategy, and some of those should remain since leaders designed them without asking for SET input. These include the Purpose, Core Values, and Vision. Each of these should be presented to the SET and explained. The remaining parts of the strategy need to be presented and explained as well; but if you choose, also discuss any ideas from the SET that could be captured for future consideration by the leaders.

Some sites designate a leader or safety professional who was present during the safety strategy workshop to present the strategy to the other members of the SET. Other sites have made this more of an open discussion of the strategy among the group who developed it and the other SET members. Many sites had not completed every aspect of their strategy, and the discussion of the preliminary strategic decisions led to some very productive discussions and a deeper understanding of the strategic issues. Remember that strategy is a living thing and can change in reaction to changing needs or simply to better strategic thinking.

STEP 3.3 SET CLARITY WORKSHOP

The next STEP for the SET is a workshop to develop clarity of mission, to outline the remaining STEPS, and to begin to deploy the safety strategy. Part of this workshop is a training session to get everyone aligned in their thinking about the basics of safety, how to improve it, how culture impacts it, and how we can create excellence within the culture. We cannot move to a higher level of excellence with our current level of thinking, so it is important that we take this first step with our SET to begin to move our thinking about safety to the next level. After we do this for the SET, we will begin to do the same thing for the entire organizational population.

We have developed a number of customized training modules and workshops for our clients in various industries and countries around the world, but they all have four common themes and modules that we will outline in this STEP. These materials can be assigned reading for your SET members, and you can have a discussion or workshop to follow or you can turn these into training modules to deliver.

1. Safety definitions
2. Accident causation and prevention
3. Understanding low-probability risks
4. The STEPS outline.

Safety Definitions

Clarity demands that the terminology we use be clearly defined and that everyone shares the same definitions so they can work collaboratively on the concepts. We discussed the definition of terms in the introductory materials, but that discussion needs to be restarted with the SET at this point. At the end of this STEP, the SET needs to be clear and aligned about what these terms mean, be able to communicate them to the rest of the population, and use them to solve safety problems.

Once these terms are clearly defined, you can begin to align thinking and actions and give everyone talking points to discuss and improve safety. As consultants, we often assess organizations and ask leaders and groups of workers to give us their definition of such basic terms as safety, accident, near miss, and so on. When the answers to these basic questions vary greatly, we find that the organization struggles to improve. When the answers are well aligned, we often find excellent performance in safety. Definition and alignment are key to clarity of purpose and effectiveness of effort. All scientists agree to the definition of key terms before they proceed with research or collaborative discussions. If we do the same with safety improvement, we progress farther and more quickly.

If your organization already has official definitions of these key terms, use them. It is never a good idea to change something that is well established or officially implemented unless the existing model is deficient or inadequate for the purpose. To determine whether your definitions are going to serve you well in this effort, compare them with the ones that follow. Remember the difference between the definition and the goal. Too many definitions of safety are really goals and not definitions, that is, "safety is not having an accident." The goal of safety is to prevent accidents, but this does not tell us what safety is or how it works. It describes a goal, but it does not prescribe how to accomplish that goal.

The defining of safety as not having accidents has another fatal flaw: it hopelessly mixes safety with luck. There are two ways to not have an accident: to be safe or to be lucky. Workers who define safety this way often excuse themselves for taking low-probability risks because they have done so many times without getting injured. Their logic is, "If safety is not getting hurt, then anything I do that doesn't get me hurt must be safe." This thinking is also reactive rather than proactive. It defines the action by the result rather than the action itself. A worker cannot know if it is safe until they see how it turns out.

A good working definition of safety has to carefully exclude the luck element and prescriptively and proactively help the worker to choose the action that will prevent the injury. The proactive and prescriptive part of safety deals with risks, not just results. Workers get injured when they fail to identify or underestimate risks. However, recognizing and appreciating risks are not enough. Workers must take precautions to prevent the risks from becoming injuries. This means that they must have a good working knowledge of how this can and has happened. They must also have a knowledge of precautions that can prevent accidents. There can be a host of reasons why a worker might not recognize a risk or fail to take a precaution, but these are the immediate issues of what constitutes safety. If your organization does not have an official definition for safety, consider using or modifying the following one.

Definition of Safety Safety is the process of identifying workplace risks and addressing them either through eliminating them, mitigating them, taking precautions to keep them from resulting in injury, or a combination of these approaches.

Or the 6-year-old explanation: safety is knowing what can hurt you, learning the things that can keep it from hurting you, and doing those things.

If the goal of safety is to prevent accidents, we should also define what an accident is. There are several safety writers and practitioners who do not like the term "accident" and advocate the use of other terms such as incident or event. One objection is that the term "accident" might suggest that the event was not preventable. We have no such issue with the term and use it to indicate that the incident or event was not intentional. To us, accidental is the antithesis of intentional or deliberate. Thus, an accidental injury (often just called an accident) is an injury that happened counter to the intentions of those involved. (See Note on Terminology later.)

We make another distinction in this definition that can be important in both understanding and preventing accidents: whether the unintentional part of the scenario was what the worker unintentionally did or whether the worker did something intentionally that unintentionally resulted in an injury. This distinction helps to identify the part of the safety definition that failed to prevent the accident. Did the worker identify the risk and fail to take a precaution or did the worker fail to see the risk and do something that was not anticipated to result in an injury? In other words, did the worker fail to take a precaution or fail to see or appreciate the risk? (Remember: We are discussing an accident that has already happened. If the risk had been eliminated or mitigated, this accident would not have happened. This is not a process to place blame or shift the burden of safety to or from managers or workers.)

Definition of an Accident (or Accidental Injury)   Accident is an unintentional action or outcome that resulted in an injury. (Not all accidents result in injury. Theoretically, anything unintended, whether a behavior you did not intend to do or a result you did not want to create, can be called an accident.)

An unintentional action is one that a worker knows will or might result in injury but does anyway for one or more reasons: lack of attention, conditional limitations, pressure, and so on. For example, a worker may know that an object is hot and will burn but accidentally comes in contact with the object. The human error thinkers tend to label these actions as "slips" or "lapses" or other category names that indicate an interruption to attention or sound thinking about the act. The fact is everyone can make a mistake, fail to pay attention, or simply go into automatic pilot mode and do something that they know they should not do. This type of accident happens, but it is not the most common type.

The other type of action is where the worker feels a level of confidence that his actions will not result in injury, but this time it does. The most common cause of this type of accident is low-probability risks. Touching a hot object is a high-probability risk. It always or almost always results in injury. These risks are obvious and can be identified with common sense or minimal experience. However, there are other risks that only result in injury one in 500 or one in 1000 times that the risk is taken. Many workers have taken such risks multiple times without injury, so common sense tells them that they will have a similar result if they do this action again. And usually they do. But one time in hundreds or thousands, the result is different. Such outcomes are difficult to predict or even explain. They are not common, so they are undetected by common sense. Experience can recognize them, but by then it is too late. This type of accident is the most common among organizations that have already been working to improve safety.

Distinguishing between these two types of accidents is important when you start addressing the underlying causes of accidents. If workers simply did something they knew they should not, you correct that a completely different way than you correct a failure to anticipate the outcome of common actions. There will be more on this when you begin to take STEPS toward Milestone 6.

Generically, there are three ways to prevent accidents, and the best approach is dependent upon understanding the distinction we just mentioned.

1. Change the conditions to eliminate or reduce the risks.
2. Change the common practice to include precautions against these low-probability risks.
3. A combination of these two approaches.

An example of these methods is how to prevent traffic accidents.

1. Can we make the conditions of driving, (i.e., the cars and roads) completely safe?
2. Do drivers need to take precautions to avoid low-probability risks?
3. Do we need to combine these two approaches: work on the cars and roads but still ask the drivers to take precautions?

Virtually all safety issues fall into these generic categories. Sometimes conditional approaches are elegant solutions to accident prevention, that is, double switches on sheers and presses to keep operator's hands out of pinch points, removing tripping and bumping hazards from walkways, certain types of machine guarding, and so on. However, sometimes it is difficult or impossible to completely remove the risk from the work conditions, that is, eliminating radiation, steam, and high voltage from nuclear power plants, eliminating steep cliffs from open-pit mines and quarries, or making the highways completely safe for drivers. In such cases, it is necessary for workers to take certain precautions to further reduce the workplace risks. In some cases, conditions can be improved, but worker precautions are also needed.

In organizations where there is good common definitions and understanding of the process needed to promote safety and prevent accidents through these three approaches, safety excellence is possible. When everyone is approaching safety from a different perspective, it is difficult to even discuss the issues much less to reach consensus. Developing these definitions and defining these generic approaches is the beginning of creating clarity of purpose and methodology for improving safety. One CEO described this as "getting all the arrows in the organization pointed in the same direction."

Note on Terminology: (Accident vs. Incident)   There is a movement away from the use of the term "accident" toward the term "incident" that is driven largely by the argument that accidental somehow implies that the event was not preventable. We believe that the term accident is perfectly acceptable and does denote or connote unpreventability. To us (and Webster), it simply means unintended. However, we strongly believe that you should not use terminology that causes confusion, lack of clarity, or misunderstanding, and if this (or any other) term does so in your organization, please use an alternative terminology. At many of the organizations we have assisted, the term "incident" is used to denote the difference between damage or loss and human injury. The official definitions tend to paint an incident as simply an event or happening, whereas an accident is an unplanned, unwanted, undesired, event resulting in injury, damage, or other types of loss. To us, this later definition is more accurate, and we have no stigma of preventability attached to it.

Defining "Near Miss" (Near Hit, Close Call, Good Catch, and Injury-Free Event)   If your organization uses or is considering using this type of data in your journey to Safety Culture Excellence, you need to make sure to define and align thinking across your entire population about exactly what this is. Occupational Safety and Health Administration (OSHA) has just recently offered their definition of a near miss. Prior to that, there was no official definition and a wide interpretation of what constitutes a near miss.

First, decide the label. Which of the many terms will be your official name for this? Second, decide if you wish to use the new OSHA definition or devise your own. We have seen organizations define a near miss as everything from an identified potential for injury to an unsafe condition to an event that did not cause injury but had the potential to do so. The latter of these is the most common, but hardly the only one. There is no right or wrong way to address this issue, but some are better than others, and whichever way you decide, it is critical to get everyone on the same page. The most common reason we have found for workers not reporting near misses is the lack of clear understanding of what the organization wants them to report.

If you chose to define a near miss as a potential or condition, you may rightly consider it a leading indicator. If you define it as an event, you really cannot consider it "leading." Leading indicators happen prior to the incident, whereas lagging indicators happen after the incident. Some view the dividing line as being between no injury and injury. In this case, they view the range of severity from minor to major. The more accurate view is that the line is between no incident and incident. The more accurate range of severity is from none to death. We encourage organizations to use the term "near miss" and to define it as an event that did not cause injury, but had the potential to have done so. We also encourage this to be viewed as additional incident data, not a leading indicator.

Defining "Preventable"   Many organizations have a statement that "all accidents are preventable," and many employees take issue with this. If the starting basis of your safety program is a point of disagreement, it is not possible to reach excellence. Workers may not be willing to work on something they think is impossible. If safety excellence or zero accidents is espoused as a goal and workers think that it is an unachievable dream, they may either not participate or may look for ways to cheat the accident data rather than to improve safety.

Theoretically, all accidents could be prevented if they were anticipated early enough. In our experience, when you ask if all accidents are preventable, people tend to think back on their own experience or instances that have been shared by coworkers or friends. Those workers and leaders who think all accidents are not preventable are generally focusing on the accident as it occurred. One person in one of our workshops relayed his own experience, "I was stopped at a traffic light. It turned green and I pulled into the intersection. When I looked to my left there was a truck running the light at about 40 mph. It hit my truck hard. Now how could I have prevented that?"

Workers and leaders who do think that all accidents are preventable tend to be thinking upstream of the accident timeline. They would determine that the time to look to the left is before you pull into the intersection. Obviously, we are not always diligent to look for potential accidents, and anyone can start into an intersection without scanning for all possible dangers. However, the formula for prevention includes early detection.

A Story to Help Make This Point   A young man volunteered to play a game with a chess master. The chess master was playing against 80 volunteers. As he came around making his moves on each board in only a few seconds, he quickly told the young man that he had not yet made his next move. The young man was amazed that the chess master could tell that with a quick glance while playing 80 games.

The master said, "Are you going to take the bishop? If you do I have you in three moves." The young man captured the master's bishop, and the master said, "Do you want to see all three moves at once or one at a time?" The young man replied, "If you do them one at a time maybe I can find a way to counter them." The master answered, "No. I have five ways to counter your three moves. When you took the bishop you lost the game!"

Question: What does a chess master do that an average chess player does not do?

Answer: Think moves ahead.

Question: What does an excellent safety performer do that an average safety performer does not do?

Answer: Think moves ahead.

Question: What does someone who thinks all accidents are potentially preventable do that someone that does not think all accidents can be prevented do differently?

Answer: Think moves ahead.

Every accident results from a chain of events. If you can see the danger developing at an early stage, you can potentially prevent the accident. If you fail to see the danger developing until too late, you may pass the point at which the accident can still be prevented.

Question: Are all accidents preventable or not?

Answer: All accidents are preventable up to a certain point in the chain of events and then are not preventable afterwards. So preventing accidents depends on thinking ahead and seeing the danger develops while still in the preventable stages.

At this point, you should capture your official definitions of these terms:

- Safety
- Accident (accidental injury)
- Near miss (or other term for this concept)
- Preventable.

These definitions should be shared in the Worker Briefing, posted on bulletin boards or communication screens, referenced in newsletter articles, and recited in meetings until they become the common definitions of the entire organizational population.

If you called us in as consultants and we asked your workers for these definitions, what would they say?

Accident Causation and Prevention

If your organization already uses a model of causation or prevention, you probably should continue to do so. It is important that you review it with your SET at this time to make sure that it is fresh on everyone's mind as you move forward. If you do not have an officially adopted model, consider the one that follows.

If safety is knowing what can hurt you, learning the things that can keep them from hurting you, and doing those things, how does an accident happen? Obviously, it can happen if you fail to recognize the things that can hurt you (risk awareness), if you do not learn the things that can keep it from hurting you (learning precautions), or if you forget or fail to do those things (taking precautions).

The first thing most organizations do in safety is risk analysis. We identify the things that can injure people in our work environment. The second thing we do is engineering and design work. There is a set of controls for safety called the "hierarchy of controls" which suggests that once you identify a problem, you ask a set of questions to find the best way to "control" the risk. The flowchart format is as follows:

Can you eliminate the risk? If yes, do so and the problem is solved. If no, go to the next question.

Each next question suggests another potential solution such as substitution, moving, putting guards up, and so on. If the condition of the risk cannot be adequately eliminated or mitigated, the next solutions are more behavioral: what precautions could people take to keep the risk from injuring them?

Organizations should do everything within reason to make the conditions of the workplace free from risks, but completely removing the risks is impossible. Do you know of any place on the planet where you can do anything you can imagine and not possibly be injured? When we hear the argument that safety should be simply removing the risks rather than working with human behavior, we tend to use an example from driving safety. Can you make the vehicles and highways so safe that it would not matter what the drivers do? Conditional safety is important and should be tried first, but the rest of safety will always be defining precautions to address the remaining risks.

Case Study: A client company asked us to help identify precautions at their site after realizing that their conditional safety efforts were not going to completely eliminate injuries. They told us about an accident where a worker tripped on the concrete edge that protruded six inches from the wall of a guardhouse at the entrance of the plant. They spent thousands of dollars removing those protrusions or expanding the walls of each guardhouse at each entrance. The next accident involved a worker who tripped on the level ground next to the guardhouse and had a more serious injury than the last tripping victim. They could not identify any conditional cause to fix that would potentially prevent a similar accident.

If you identify a risk and cannot completely eliminate it with conditional interventions, the next step in safety is to identify a precaution or precautions that can reduce the probability of injuries around that risk. These precautions can include such issues as ergonomics, body mechanics, where you position yourself in relation to the risk, communication among workers, housekeeping, tool and equipment use, and personal protective equipment (PPE). Later in the STEPS process, you will learn how to perform an analysis to determine which precaution could impact the largest percent of your accidents and/or near misses. This analysis will help you develop the ultimate STEPS to address safety issues and to reach more excellent levels of safety performance.

Precautions can fail to prevent accidents for two reasons: (1) the precaution is inadequate to address the risk; or (2) the precaution is not taken for some reason.

Inadequate precautions are usually well-intentioned actions that either under-estimate the risk or simply do not go far enough in protecting from the risk. Rules and procedures can sometimes be such well-intentioned but inadequate precautions. Workers who think that the rules and procedures will automatically keep them safe are often surprised when they fail to do so. Rules and procedures are often minimum standards of safety and should never replace risk analysis on the job. PPE can also offer a level of protection that is inadequate for some risks but can create a false sense of security. The accidents we review are full of examples where workers fol-lowed the rules and procedures and still got injured, where workers did a job just as they were trained to do it and got injured, and where workers were wearing the prescribed PPE and still got injured. It is important to remember that taking the wrong precaution or an inadequate precaution is no guarantee of safety.

Precautions are sometimes not taken. Workers sometimes simply forget to take a precaution. Sometimes they deliberately bypass a precaution to save time when they feel rushed or are being pressured to increase production. Sometimes a precau-tion simply has not been taken often enough to become automatic or habitual. Sometimes the culture has not accepted a precaution, and taking the precaution has not become common practice. Cultures can actually discourage some practices that seem foreign or a waste of time. Failing to share the rationale for the precaution can cause such misunderstandings or resistance within the culture. The regular taking of precautions is greatly reinforced when the precautions become habitual and cultural norms. This means that workers take the precaution even when they do not make a conscious decision to do so and that they probably get approval from their fellow workers when they do.

Understanding Low-Probability Risks

There is another reason for not taking precautions that we find to be a significant problem in many organizations. This reason is the failure to recognize or the ten-dency to underestimate low-probability risks. Probability is a concept that is critical to understanding safety but is not accurately understood by most workers. Once an organization has a reasonably mature safety effort in place, most of the remaining accidents usually involve low-probability risks. We regularly find this issue to be the key to moving to the next level of excellence in organizations that are already have achieved improved safety performance.

To understand the impact of low-probability risks, think of how most workers approach safety. Organizations usually have safety training as a part of new-employee orientation and ongoing training as required by regulatory agencies and/or company policy. But, on the job, how do workers make day-to-day decisions about safety issues. They use not only their training but also their common sense and experience. There is a hot debate about the term "common sense" and if there really is such a thing and exactly what it is. But most workers use that term and to them it means obvious things you should be able to figure out. You should know what will happen if you stick your hand in ingoing gears or your screwdriver in an electrical socket. If you cannot figure that out, you do not have "common sense."

When a worker has been on the job for a period of time, he or she has personal experiences and sees things happen to others that they did not predict with their common sense. They tend to call these things "experience." You may hear longer tenured employees saying things like, "When you have been around here as long as I have, you don't. . . ," or, "If you have ever had to break into that line before you know what can happen when. . . ." Experience enhances and expands commonsense thinking and tends to make workers safer if they learn lessons and take precautions.

The problem with common sense and experience is that they usually do not detect low-probability risks very well until it is too late. The term "common" in common sense also implies regular and predictable. Things that happen regularly (every time or most of the time) can be predicted with common sense. Things that happen less regularly can sometimes be predicted or implied from common sense or similar experience or the experience of others, but when the probability or regularity of the occurrence becomes low enough, it is not predictable. Workers do not learn about nonpredictable or low-probability risks until after they have been injured. Even workers who receive only minor injuries do not take precautions until after a more serious injury occurs.

The problem of low-probability risks is compounded if workers think of safety as "not getting hurt." The most common consequence of taking a low-probability risk is not getting hurt. If you have taken a risk multiple times without getting hurt, your brain will predict that this pattern will continue until your experience proves it wrong. To become excellent in safety, you have to think of safety as what you do to prevent or lower the probability of an accident and not simply as either having or not having an accident.

We call this concept the difference between reactive and proactive safety. Most organizations react to accidents. Doing so can improve safety, but only to a limit. That limit is defined by two major factors: the probability of the remaining risks and the statistical significance of accident data. When you have a lot of accidents, you have a lot of accident data. You can analyze that data and often find commonalities and focus on high-impact preventions. As that data diminishes, it ceases to have trends and commonalities and is less rich in potential solutions. As the organization pursues accident prevention, it usually works on the highest probability accidents first and works down the probability scale. By the time the organization is pretty good in safety, the main accidents left are low-probability accidents, and the data describing those few accidents are not rich and full of obvious solutions. At this juncture, the organization needs to become proactive to reach the next level of excellence in its safety performance.

There are two training modules we have used so often to illustrate these concepts that we have made them into short training videos. Rather than describing them here, we would like to offer them to you at no charge and recommend that you either use them in your training or use them as models to design your own training modules. They are entitled "Probability" and "Cliff Analogy" and can be found at http://www.proactsafety.com/insights/videos.

The STEPS Outline

The final section of this training and workshop should be to show and discuss the remaining STEPS that will need to be taken or bypassed by the SET on the way to Safety Culture Excellence. Use the flowchart in Figure 9 in "Making the Decision to Pursue Safety Culture Excellence" to illustrate. Part of creating clarity is to have a roadmap of where you want to go and knowing the milestones you will pass along the way. If you are unsure of your path, you are lost. When we feel lost, our tendency is to go backwards and find a familiar place and reestablish our way forward. Roadmaps avoid that loss of direction and clarity of purpose.

As in all STEPS, your organization may have already addressed the issues in some of the STEPS, and you may choose to skip them entirely or to simply look for ideas to improve your existing efforts. That is perfectly in accord with this methodology although we do recommend that you take opportunities to recognize your own strengths and accomplishments along the way. Being able to skip a STEP means that you have something good in place and that may be a good reason to thank those who accomplished it or point it out to create pride and momentum in your efforts. Excellence is difficult, but it is also contagious, and you should take opportunities to expose your people to it and let them get infected.

Application of STEP 3.2: At the end of this training and workshop, the SET should be clear on the terminology of safety and the methods to prevent accidents. They should use this knowledge to do their jobs in the SET, and they should share it with their fellow workers to help create clarity in safety efforts throughout the organization. These definitions and methods, like everything else in the STEPS program, may change as issues evolve and your knowledge grows. If they do change, the change needs to be communicated and shared with everyone to keep clarity and focus in safety excellence.

STEP 3.4   STEPS EMPLOYEE BRIEFING(S)

Now that you have your SET informed about the safety strategy and the remaining STEPS to Safety Culture Excellence, it is important to get everyone else in the organization informed and on board to the greatest extent possible. No one will have enough information at this point to fully understand what this process will entail or accurately envision their role, but they can begin to see the rationale and strategies to be used. For these reasons, we recommend that you begin this STEP by assigning a person or group of people to prepare and deliver a briefing to everyone in the organization. In smaller organizations, this might be accomplished in one all-hands meeting. In larger organizations, it might entail a roadshow or a set of training sessions that accommodate shifts, departments, and/or other logistical issues.

The person or persons who deliver this briefing should be well informed on the STEPS methodologies. This means that they should be a member of the SET and have, at a minimum, completed reading this book and/or received formal training. Such a basis of knowledge is needed to deliver the material, but it is even more critical in answering any questions that arise during the briefings. Select a person who is an effective and credible communicator. Do not be too concerned with formal platform skills and do not make your selection based strictly on education or eloquence. Select someone who can "talk to the troops," answer questions, and has the authority to leave the impression that this is officially what we are going to do. The goals of these briefings are as follows:

- To make sure everyone knows the rationale for pursuing safety excellence
- To inform everyone on the basic concepts of STEPS
- To enlist support and buy-in for the process
- To dispel any misconceptions about the process
- To answer questions that may arise.

The length and scheduling of these briefings are important elements in achieving these goals. If these briefings are too short, they may not convey enough information, adequately answer questions, or allow enough time to get all questions answered. Rushing can also send the message that this process is not important. If the sessions are too long, workers may let their attention drift or get overloaded with information. Either way, the wrong length can render the briefings ineffective.

The scheduling of the briefings can also send a message about the importance of STEPS. If they are held after shifts when workers are tired, this could send the message that they are not important enough to interrupt work and that there is no concern that it is being delivered to workers in poor condition to be taught. Every organization is unique, and there is no universally perfect time and length, but we recommend that every organization review the guidelines later and customize these briefings to best fit their needs while being true to the principles of effectiveness.

We recommend that the briefing be scheduled so that it sends the message that this is important enough to take priority over everyday matters. The way that is accomplished can vary greatly within organizations and, in some instances, may be difficult or impossible. However, if the attempt is made and the priority is expressed in the briefing, the workers should begin to get the message. We also recommend that these briefings only last about 30–45 minutes. That is enough time to convey sufficient information and answer questions without losing attention or overloading workers. It is also a time that can often be taken by bringing a shift in early and by holding the other shift over to cover. Always try to teach workers coming on rather than going off shift. This maximizes freshness and mental ability to absorb information.

We recommend that the briefing contain the following main points:

1. *The organization's rationale and objectives for wanting to achieve safety excellence.* Some organizations get from bad to good in safety and chose to

move their attention away from safety to other priorities. For a select few, there is no "good enough" in safety, and they chose to continue to strive for better and better safety performance. Some are driven by a desire for excellence and others by client demands, empathy for people, or competitive advantage. It is a good idea at this juncture to deeply question your own organizational priorities and define your true reason(s) for wanting to be better in safety. It is going to be difficult to sell your workers on this endeavor if you cannot sell yourself. Questioning and defining your rationale will also help you to articulate it and to develop the right dialogue to guide and direct your efforts. Defining your rationale may or may not give rise to slogans or sayings that can be used to communicate the importance of this effort to the organization. Answering the proverbial "What's in it for me?" question is not a bad way to begin. Envision an excellent safety culture producing excellent performance and ask what this would mean for you, your employees, their families, your stockholders, your clients, and so on. Help everyone begin with the end in mind so they can clearly see the destination and help you get there.

2. *Share the safety strategy.* The safety strategy that was developed in the workshop with managers should be outlined and shared with the workers. It is not necessary at this juncture to detail every aspect of the strategy, but each major section should be mentioned and an overview of the details provided. The site population should leave these briefings with the feeling that the leaders have a plan and that the plan is complete, but flexible. In addition to sharing the strategy, you might consider sharing how the strategy was developed and exactly who was involved.

3. *Why the organization chose the STEPS methodology.* Obviously, we believe that the STEPS process is the best way to reach Safety Culture Excellence or we would have recommended another course of action. But, why do *you* believe that it is the best way for your organization? How does it fit in with your mission, vision, values, and/or other programs and initiatives? How is it a good fit for your people and existing culture? Does it fit your budget, timeline, goals and objectives, and so on? What other methods did you compare and contrast and why did this one win? Explaining the process of selection, the criteria for selection, the people involved in the selection process, or other details can help everyone better understand that this is the best approach and that the decision to do STEPS was a sound one. Your objective is to get buy-in. The more people know about the range of options and the advantages of the chosen methodology, the better you can accomplish this goal. This is also the case when you share your safety strategy and the rationale behind it. Mention how it fits in with your vision and values for safety excellence and mention the people who contributed to its development. Remind everyone that such strategies are flexible and that new ideas or methods are always welcome. Be careful to set realistic expectations about how suggestions will be considered and evaluated and how long that process will take. Do not create the false illusion that all suggestions will immediately be adopted and implemented.

   Caution: If your organization has a strong sense of "flavor-of-the-month" programs, you might want to posture this as an evolution of existing efforts or simply as continuous improvement rather than giving it too much of a "program" flavor. Do not be deceptive, but present your strategy and use of STEPS in the best light to get buy-in and start toward quick wins to build momentum. Do not let terminology slow your progress.

4. *Stressing the point that excellence is not necessarily perfection.* STEPS is designed to help a safety culture methodically reach its personal best. Most cultures are not capable of absolute perfection and setting that as a goal can be demoralizing. The first goal of STEPS is to start moving in the right direction. Learning how to take a step toward the goal is the basic capability necessary to reaching excellence. If you truly realize that excellence is a journey, then the measure is not your current location, but your progress and your ability to make progress. The goal of this section is to set challenging but realistic expectations for the STEPS process, to challenge the imagination without overloading it, and to start workers on an optimistic journey without making them feel pressured or rushed to accomplish the impossible in an unrealistic time. If workers change their tolerance for their own and the organization's safety performance and begin to be dissatisfied with anything less than excellence, you are on the right road.

5. *An overview of the STEPS remaining.* When people go on a journey and get lost, their first tendency is to backtrack to try to find a familiar and known location. The same tendencies will take hold in STEPS if workers or the organization lose their way. For this reason, it is wise to provide everyone with map of where the journey will take them. It is not possible to cover every aspect or detail of the process in a meeting of this length and scope, but an overview of the main points can greatly help to win over hearts and minds and begin to map out the journey ahead. A list of the milestones and the STEPS leading to each one can be shared in this meeting and posted for future reference. Copies can be handed out at the meeting or made available. STEPS can be marked off as completed to show progress. Keeping the path ahead fresh in the minds of the workers is a key part of maintaining clarity and motivating participation in the process. A good visual to use during this presentation is a graphic of the milestones in the STEPS process (see Figure 3.1).

Some organizations who helped to pioneer the STEPS process actually asked workers and managers to read a brief article about the process or had the article read in meetings prior to holding this briefing. Others published the article in newsletters or posted it for review. One organization assigned supervisors to read the article and relay the key ideas to workers in shift start-up meetings a week before holding the kickoff briefings. Whether or not you require people to read or discuss ideas prior to the briefings is an individual decision. Proponents found that it started people thinking about the topics and gave them a boost in the briefings. Opponents claimed that it created misconceptions, fed the rumor mill, and caused the briefings to address the problems before jumping into the real content. If you would like to use the article as mentioned or make it available to your associates, it can be read or copied from the website http://www.proactsafety.com. If you do not leave any other

Figure 3.1   Milestones on the STEPS Pathway™.

impression, make sure that workers know that the organization is going to be focusing on one issue at a time, addressing it, and moving on to another issue.

6. *Inform everyone of the personnel involved directly in STEPS*:

(a) Make everyone aware of who the members are of your SET and/or your sponsor or facilitator for the process. List these people and their positions and titles during the training. If not everyone knows who these folks are, state their job titles and the division, area, department, and so on where they work. The idea is to demonstrate that the SET has a representative from each major area or division so that workers start to gain confidence that their issues are understood and addressed. It is also important that everyone knows someone to contact if they have questions or suggestions for the process. If you have an organizational designated sponsor or facilitator, make that person's name and role also publicized in the briefing.

(b) *Suggestion*: You should extend an invitation to everyone to make suggestions, offer ideas, or comment on the process to those involved. The process does not just belong to a few team members; it belongs to the whole organization, and you should invite everyone to participate. Many organizations invite people who would like to be considered for future roles in SETs or other positions to make their wishes known also. Again, this volunteering should be done privately rather than in public in most instances. For example, ask volunteers to tell their SET member or supervisor rather than asking them to stand or come forward during the briefing. This briefing is also a good place to announce any plans for communicating STEPS information. If you are going to utilize existing media such as bulletin boards, newsletters, information screens, and so on, or if you are going to create new media specifically to communicate STEPS data, this briefing is a good place to make such plans known to everyone and start the anticipation of such data. Some STEPS processes designate communication specialists for the special tasks of designing the messages or managing the media for communications. If you have done so, make sure that these people get recognition or use the briefing as an opportunity to get volunteers.

Application of STEP 3.3: After the employee briefing session(s), workers should know that the organization is seeking to pursue safety excellence and why it thinks that it will be beneficial. They should know that leaders have a strategy, the starting place has been assessed, and progress will be measured. They should know that the organization will target specific things to improve, work to improve them, and move on to other targets. They should have a roadmap of the STEPS to refer to and know the people on the SET who represent them and to whom they can go with suggestions and ideas. They should know where more information will appear on the progression toward Safety Culture Excellence. It is impractical to assume that they will understand or remember every detail about the process. This is a "kickoff" of STEPS, and it is important that everyone be aware of it, begin to get comfortable with it, and better understand the process.

CLIMATE

It wasn't raining when Noah built the Ark.
—Howard Rough

Now that you have a safety strategy and the structure to improve safety, clarity of purpose and definition, and have introduced your plan to your organization or site population, you are ready to start working on the climate. There is a belief among some that culture is the byproduct of climate and that the climate is the ultimate control point for improving the culture. Whether or not you subscribe to that philosophy, the organizational climate can either encourage or stifle the growth of excellence in the culture. Like a plant, a culture needs the right atmosphere in which to grow to its fullest potential. Climate is largely the purview of management, but we recommend allowing the Safety Excellence Team (SET) to have some voice in how it can be improved since the connection of the levels of the organization also impact the culture. The model in Figure 4.1 shows the three most common levels in most organizations forming the foundation and roof of a structure (leaders, supervisors, and the workforce). Connecting these levels are the four pillars labeled Commitment, Caring, Cooperation, and Coaching. The enhancement of these four elements should be the focus of reaching this milestone. Enhancing the organization's capabilities and performance of these four issues will create the kind of climate for Safety Culture Excellence.

Creating the right climate for growing an excellent safety culture involves four STEPS. As in all STEPS, you may have already accomplished them or have done work toward accomplishing them. They can be skipped or only partially addressed in these cases. But we recommend that you read through each step even if you feel you have already addressed it to make sure you have not missed an important part or that you do not miss a further opportunity to improve. We recommend that you either assign leaders to read and discuss this milestone or hold a workshop for the SET to address each of the four STEPS you select. From this workshop could come recommendations for organizational changes or training for some, or all, of the workforce to enhance these elements.

STEPS to Safety Culture Excellence^{SM}, First Edition. Terry L. Mathis and Shawn M. Galloway.
© 2013 John Wiley & Sons, Inc. Published 2013 by John Wiley & Sons, Inc.

Figure 4.1   The four pillars of Safety Climate™.

At this juncture, some organizations ask the senior leaders to address the four STEPS of Climate while the SET moves on to address the STEPS of Chemistry in the next milestone. Such a division is not necessary but has been successful in a number of organizations. This option is the reason that Climate and Chemistry are parallel rather than linear in the flowchart.

Goals: To create a commitment to safety excellence

To drive the safety efforts by caring about each other

To establish a basis and encouragement for cooperative efforts among the various levels

To establish coaching as the method of helping each other to improve performance

To train everyone in the skills of coaching safety

To create an environment conducive to growing an excellent safety culture

Methods: SET workshop

Declaring commitment to safety excellence

Changing the aim of safety from statistics to people

Creating forums and communication tools to encourage cooperation

Training

STEPS: 4.1 Commitment

4.2 Caring

4.3 Cooperation

4.4 Coaching

STEP 4.1   COMMITMENT

To grow an excellent safety culture, there needs to be commitment to doing so at every level in the organization. This commitment should begin at the top of the organizational chart to be sustainable at the other levels. It is not easy or sometimes possible to remain committed if your boss is not. Safety excellence is an executive commitment before it becomes a management commitment. It is a management commitment before it becomes a supervisory or worker commitment. To successfully impact every level in the organization, safety excellence must be sponsored at the right level in the organization. Safety culture does not necessarily start with leadership, but commitment must.

Commitment to safety improvement does not replace or preclude commitment to other goals. The reality of most organizations is that safety is not their primary purpose for existence. They are in business to make a product or provide a service, or they are an agency of government or a nonprofit with a specific mission. Safety excellence does not replace that goal, but it modifies it to include doing the job as safely as possible and continuously improving how safely it is possible to do the job. Organizations that adopt slogans that indicate that safety is a higher priority than the very reason for the organization to exist run the risk of actually undermining their own safety efforts. Workers seldom believe that safety is really the top priority. When leaders say it is, they may undermine their own credibility. The reality in most organizations is that they want safe production, not safety at any cost. If that is your reality, why not communicate it accurately?

Commitment is also not accepting any level of "good enough" that is less than excellent. It is not withdrawing effort from safety when you reach levels of acceptable performance. It is not stopping with momentary success. It is realizing that excellence is a journey and not a destination. Getting to excellence requires commitment, and staying there requires *ongoing* commitment. You cannot merely say it but must provide time and resources and maintain the priority or value of safety improvement in the face of competing priorities or values. You must do so even when levels of current performance do not demand it.

Such commitment requires leaders, managers, and supervisors to accept that safety excellence is not just an altruistic wish but also a good business decision. In many industries, safety excellence is a condition of doing business with other organizations who are also striving for safety excellence. Even if there is not pressure from clients or suppliers to perform in safety, such performance is ultimately a competitive advantage. Almost every client organization we have worked with reported that as safety improved, other aspects of business improved proportionately. One organization told us that their quality problems almost disappeared when their accident rate approached zero. Another organization told us that union grievances diminished proportionately with accidents, and hundreds of sites told stories of how housekeeping and employee job satisfaction improved along with safety. One client bought a factory and wanted to make it into a model site within their company. They found that they could not recruit the kind of talent they wanted because the site had a history of safety problems, and they had to address safety before they could hire the caliber of employees to help them meet their goals.

We are convinced that good safety is good business, and excellent safety is excellent business. Can your leaders come to the same conclusions and develop the same commitment? Leaders who are not convinced of this will often subordinate safety to other (in their eyes, more important) priorities or will lack consistency in their approach.

Workers often appreciate the importance of safety already since they are the victims of its failures. However, workers can be diverted from safety efforts by leaders and supervisors who have other priorities or that send inconsistent messages. Workers' levels of commitment to safety excellence can also suffer if they do not believe that further improvement is possible, that it would not be supported, or that priorities will change. The right level of commitment at all levels must come from true conversion at the top, consistent communication between levels, and knowledge of how to proceed at all levels.

Commitment must be attained, but also maintained. The level of commitment will be tried and tested many times before it becomes accepted. Every time a key decision pits safety against other priorities or values, the eyes of the organization will be on the decision maker to see if the commitment is still there. These decisions are the indicators of commitment to the general population. There is a popular belief in many organizations that the true priorities are revealed in the decisions and in the budget, not just in the leadership rhetoric. It becomes crucial that not only leaders demonstrate their commitment but that the results of decisions and expenditures that indicate this commitment are meticulously communicated to the general population and that any misinformation about such issues be corrected.

Where trust levels are low and building commitment quickly is important, consider making a formal statement of the commitment to safety excellence in a charter for the project. This document can contain a mission statement, goals, values, methods, and so on as well as time and resources that are going be pledged to the effort for a specified period of time. If such a charter is developed, it must also be circulated, explained, and allowed to be questioned by the population. Charters can also be helpful at locations where there is regular turnover of site leaders, and efforts can be lost in transitions of management or leadership.

Case Study: We had the opportunity to interview the senior safety officers of several large organizations that regularly hire contractors and set very high safety standards for those contractors. We asked each of them the question, "Why do you require a high level of safety excellence from your contractors?" The replies from each were amazingly uniform. Their first response was simply that they had reached a level of safety excellence in their own organizations and did not want to compromise that by mixing their excellent safety performing employees with contractors that were poorly performing in safety. Then each respondent got a serious and reflective look and added, "By the way, you know if these contractors can't manage safety, they probably can't manage other important factors like quality." Many leaders view the safety performance of an organization as much more than safety. They view it as a window into how well the organization manages itself in general.

STEP 4.2 CARING

There is an old saying that no one cares how much you know unless they know how much you care. Obviously, either you or someone in your organization cares about safety or you would not be exploring ways to improve it. But the rest of the issue has to do with *why* you care about safety and if you really care about people. In the marketing STEP, we discussed how making safety personal can get the heart involved and elicit discretionary effort. This kind of effort is essential to reaching the highest levels of excellence, and anything less will suboptimize the results.

A good first move toward this goal is to modify the way the organization publicizes and responds to lagging indicators in safety. When the organization stresses reaching certain levels in frequency and severity rates, it can possibly send the message that the numbers are more important than the people. If accident data are communicated through stories with sincere human interest and workers are urged to prevent future injuries to alleviate human pain and suffering rather than make the numbers look good, workers tend to get more involved.

One problem with the standard numbers in safety data is that workers often do not really understand them or see how their performance directly impacts them. We have interviewed over 7000 workers during assessments at client sites over the past 10 years, and less than 5% of those interviewed can tell us how an Occupational Safety and Health Administration (OSHA) recordable rate is calculated. When we ask workers how their performance impacts these numbers, the overwhelming answer is simply, "Well I guess if I don't get hurt I help the numbers and if I do get hurt I make the numbers worse." This thinking reinforces the definition of safety as "not getting hurt" and justifies anything that does not result in an accident. It can also drive the concept that management cares more about the numbers than the people.

Another way to show people that you care is to listen to them. Mark Sanborn said that "Listening is empowering people by taking them seriously." In most organizations we have worked with, the workers have a lot of good ideas and observations that could be used to improve safety. Often the difference between excellent results and pretty good results comes from listening to the experts: the ones who do the job. In almost every case, workers who are listened to and taken seriously feel more cared for than those whose voices fall on deaf ears.

One factor that can quickly destroy the perception that you have a caring organization is how you use discipline for safety. If discipline is automatically and always used when accidents happen, it can send the message that the organization views people as the problem rather than the customer of the safety program. It can also send the message that the organization would rather fix the blame than fix the problems. Interestingly, in our practice, we have found that workers react almost equally to the underuse of discipline as to the overuse. If individuals can blatantly violate the safety rules without a penalty, workers do not think the organization is doing its job in safety. We think that discipline has a place in safety and should be used correctly and judiciously. We recommend that three things should prompt the use of discipline: willful violations, flagrant violations, and repeated violations.

Good workers can occasionally err. To punish well-intentioned workers who are usually safe is potentially counterproductive as well as uncaring.

Become sensitive to the tone of safety communication and see if it is about people or about numbers. Is safety aimed at blaming people or accomplished by listening to people? Is the primary goal of safety to avoid pain and suffering or is it about making the organization look good on paper? If you discover ways in which you are sending the wrong message or not showing how much you care, start to communicate the message of caring and stay on message. Utilize your perception surveys and interviews to test the pulse of the organization and see if your message is really making a difference. Set an example of caring in how you react to accidents and how you treat accident victims.

Case Study: We were working with a large organization several years ago and one of the clients who was attending our training told us this story. He said that he was required to hold a regular safety meeting with his direct reports, and our training was scheduled to end at exactly the same time his meeting was scheduled to begin. Because time was short, he skipped his customary trip to his office for his laptop and projector that he normally used to share the latest round of typical lagging indicators in safety with his team. Instead of talking about the numbers, he took the occasion to tell his team members how much he cared about them and how that care had led him to volunteer for additional safety training that he sincerely hoped would help to improve safety and prevent the terrible injuries that had happened recently. At the end of his meeting, every member of his team lined up to shake his hand and tell him that this was the best safety meeting they had attended in all their years at the facility.

He was a bit surprised and he responded, "But I always tell you I care and then I usually show you the numbers." They responded that the message they received from that approach was that he cared about the numbers. It was not his intent, but that was the way his message was received. It is not hard to be misunderstood, and it is important to occasionally stress that you care.

STEP 4.3   COOPERATION

If you are going to achieve Safety Culture Excellence, you must graduate from the mindset that safety is about management getting the workforce to comply, to think before they act, and to pay attention. Excellent safety is always a cooperative effort between all levels in the organization. This is one reason we recommend that your SET be made up of representatives from all levels and areas of the organization. Such a team can be the beginning of a cooperative effort that incorporates the voices and points of view of everyone through their representatives to move toward excellence in safety. If you are going to improve the culture, you need to include all levels and factions of the culture in that effort.

Many social scientists have suggested that empathy is the beginning of human cooperation. Whether or not you subscribe to these theories, it makes sense that if safety is centered on empathy (caring for individuals rather than making numbers look good), it will be easier to get more people to cooperate. Also, if the effort involves emotions that touch the heart, the level of effort will increase. Most cooperative efforts begin with a call to action. The beginning of STEPS should be such a call, and the rationale for the call should be caring as well as cooperating.

Language is also central to the ability to cooperate. This is another reason why we think it important to develop and share definitions of the key terminology of safety. When people share terminology, they develop talking points. These talking points facilitate and encourage discussions. Discussions tend to build differences into points of consensus, and such points are at the core of culture. By getting everyone on the same page (literally and figuratively) in safety, you can begin to build cultural norms of safety that can become common practice and perpetuate excellent performance.

Another argument for cooperative efforts is the idea of synergy. Synergy is the idea that the whole can be greater than the sum of the parts if the parts synergize together. A great example of this concept is when a group is brainstorming and one idea gives growth to other ideas that might never have surfaced without the additional point of view. Even when one group solves the problems, another group may have infused ideas that expanded thinking and fostered creativity and problem solving. When everyone is involved in safety, the ideas of how to improve are almost always better and more plentiful.

As with caring, cooperation can also be damaged by improper use of discipline. When the attention of the organization turns from fixing problems to fixing blame, the harmony of the group suffers. Blame fosters defensiveness, and defensiveness tends to close down effective communication and stop interaction. Punishment that is perceived to be justified, deserved, and dealt out with caring rather than vindictiveness does not do as much damage to cooperation. But the overuse or misuse of punishment for safety issues can bring cooperation to a standstill. This does not mean that discipline should stop; it means that it should be used justly.

There is much debate in the social sciences as to whether humans tend to cooperate or to look out for their own self-interest first. There are experts on both sides of the issue, and there is not always consensus. But ongoing experimentation on such issues as the "prisoner's dilemma" tend to indicate that humans do have a tendency to want interaction and cooperation and will seek it out and use it under at least some circumstances. The challenge for STEPS is to find the right circumstances, language, synergy, empathy, and harmony to get organizational teamwork working for safety excellence. In our experience, what works for one organization is not always a blueprint for how to make it work in another. But we do find these commonalities in most successful efforts. The other commonality for success is an organization that cares deeply enough to diligently seek the best way to cooperatively succeed.

Case Study: We were working with a site that had two different mills each with its own union. The history of the site was one of contention between management and unions and between the two unions. The company sent a new manager and safety manager and gave them an ultimatum to either improve the sites quality and safety or the company would close the site. The new manager was a newly promoted engineer in his first site management position, and the safety manager was from a different part of the country and had an "accent." The workers from both mills labeled the manager an amateur and the safety manager a "foreigner" and did not cooperate well with their suggestions. We recommended that the safety manager designate a safety representative from each site (which involved each union) who had good credibility with the mill workers. We suggested that he communicate his ideas through them until he grew in credibility. The two reps selected were a former union president from one mill and a former union vice president from the other mill. Almost instantly communication began, and the new site manager and safety manager quickly learned through their reps about the workplace challenges to quality and safety. The safety reps told the workers that they respected the new managers and trusted that they were working hard to save the mills. Attitudes began to change, and information and ideas began to really flow. Soon, the stigma was gone from the new managers, and the cooperation improved between management and unions and between the two unions. After a year of cooperative efforts, the site had moved from the worst performing in the division to the second best. That was over a decade ago, and the site is still open for business.

STEP 4.4   COACHING

Coaching, the ability to help another human being perform better, is critical to Safety Culture Excellence. Unless individuals can improve, the culture and the organization cannot improve. Excellent cultures do not come strictly from great leadership; they come from leaders who are leading great people. The most successful leaders realize that success is not just about them and how they perform each day. It is about the people they lead and how they perform. Leaders who focus on developing the talents and capabilities of their people are the greatest leaders and usually have the greatest organizations.

But leaders are not the only ones who can learn to coach and help others improve. Supervisors and workers can also become coaches and help improve safety performance. Coaching is a skill that can be taught, but it must also be managed to realize its full potential. The organization is encouraged to adopt and teach a model of coaching. Coaching should become a required training class and a part of everyone's job description (roles, responsibilities, and results [SRRR]). Coaching should be a category on every performance appraisal and job review and a topic of regular conversation. There is no skill more basic for excellence than coaching.

The basis of most coaching models involves observation and feedback. If done properly, coaching does not require extra time or structure. Almost everyone sees others in the workplace and talks to them on occasion. Supervisors and managers see what their direct reports are doing, and they discuss it. Workers see what their fellow employees are doing, and they talk to each other. Coaching is not another communication; it is a better and more impactful form of normal communication. Although many coaching models suggest specific patterns of communication, it is important not to script coaches, not to put artificial words in their mouths. Coaching should become a normal mode of communication and should be personal and sincere. It is hard to sound sincere saying someone else's words.

Figure 4.2   Four vital questions for managing performance.

Coaching is a form of, and part of, performance management, and following a basic model from that field can be helpful. The model we chose, Figure 4.2, is one that helps create a visual model for understanding how to help another person improve their performance. There are four parts to the model and each can be expressed as a label or a question.

1. *Behavioral targets.* What do you want people to do? One of the most common mistakes in coaching is failing to set improvement targets. If you simply watch someone and evaluate their safety performance, it becomes an issue of your opinion versus theirs. If you set improvement targets, it focuses efforts and empowers meaningful feedback. It is not, "How are you performing?" It becomes, "Did you do the targeted behavior or not?" This kind of target for improvement creates coaching moments that are natural. The old artificial "me evaluating you" is a thing of the past.

2. *Communications (level of expectations).* How do you communicate it to them? Once you decide on behavioral improvement targets, you must communicate them. It is important to think of communication as a process, not an event. Telling someone one time will seldom do the job. It is almost always better to enlist them in selecting or agreeing to targets than simply dictating commands. If they agree to work on an improvement target or even help to select one to work on, the level of effort is greater, and coaching is more natural, expected, and appreciated. Constant and repetitive communication can serve not only as a reminder but also to set a level of expectation that this is not going away and really needs to be accomplished. Be careful not the make repetition seem demeaning or remedial. Keep the tone of a coach and not a parent or critic.

3. *Consequence for performance.* What happens if they do it? This point is a strength of coaching and a weakness of many management styles. Managers

may remain silent when they get what they asked for. Coaches see that this as an opportunity for positive reinforcement and a way to build on strengths and to improve both performance and relationships. Good coaches give thumbs-ups and "atta-boys" and "good jobs" on a regular basis. Coaches try to never let good performance go unreinforced.

4. *Consequences for nonperformance.* What happens if they do not do it? Managers see nonperformance as a call for criticism, correction, or even punishment. Coaches see it as an opportunity to find out "why?" A good coach realizes that people do things for a reason and that, if you do not change the reason, you might not change the behavior. Coaches also realize that forced change is usually temporary, and they seek to help workers change themselves rather than be changed by authority and command.

Teaching and practicing this or alternative models of performance coaching can accomplish some really good benefits for the organization:

- It can move the organization away from command and control to a more cooperative and caring management style.
- It can teach a more effective model of interpersonal communication.
- It can create a mindset of continuous improvement as an organizational norm.
- It can foster an atmosphere of building on strengths and positive reinforcement.
- It can develop a spirit of teamwork to win the battle against accidents.
- It can standardize the way everyone manages and supervises and create good talking points to help people to help each other increase their coaching effectiveness.
- It can eliminate a lot of negatives connected to alternative ways of managing.

Adopting coaching as a skill and requirement can greatly facilitate the other aspects of creating a conducive climate for Safety Culture Excellence. Coaching begins with a commitment to improve. It is a more caring way to improve performance than most other management models. It is more cooperative and less confrontational. It can be the linchpin of the effort to improve the organizational climate.

Case Study: A safety professional we had worked with in the past called us into her new company and asked for our help. Her new company had hired her to produce a quick improvement in safety and then to put solid programs in place to ensure continued safety performance. She had done a safety intervention at her last company, and her first thought was to duplicate it here. After an assessment, we collectively decided that what had worked at her last company was not a good fit at the new company. The assessment also revealed that the average supervisor was not actively involved in safety and did not feel that they had been trained to supervise safety effectively. As an intermediate step to start the new company toward quick improvement, we conducted a 4-hour safety coaching class for all managers and supervisors. We also prescribed a way to get them to start using their new coaching skills and to hold them accountable for doing so. The company had a 55% reduction in frequency rates and an even greater reduction in severity during the next 6 months and has not quit producing further reductions since.

Milestone 4 Application: Consider reading and discussing these STEPS or holding a workshop on each of them. This should move leaders to better realize how they set the tone in the organization and either encourage or discourage the safety culture to move toward excellence. It is also critical for leaders to realize that once they have adopted these principles and started to improve them, it is important to communicate their efforts and to publicize their successes. It is not just what leaders think and feel about the Climate that is important; it is what they say and do. If leadership actions to improve Climate are not effectively communicated and demonstrated to the SET and the rest of the organizational population, they will not be as effective as they could be. A Climate can be felt and detected and judged. If workers do not see what leaders are doing, they will feel, detect, and judge the Climate by the other factors that make it up. Climate improvement must be felt by the workers if it is to encourage the safety culture to grow toward excellence.

CHEMISTRY

> *Even the richest soil, if left uncultivated will produce the rankest weeds.*
> —*Leonardo da Vinci*

The goal of this chapter is to offer an in-depth discussion of each of these elements and allow you to take STEPS, if needed, to improve any or all elements you deem in need of improvement as you progress toward Safety Culture Excellence. If you followed our admonition in the introduction to read the entire book before beginning your process, this will not be a surprise. Just as a growing plant needs the right elements in the soil for maximum growth, a safety culture needs the right elements in the organization to maximize its true potential for excellence. We not only recommend that you visit these elements during the STEPS toward your goal, but that you periodically reassess them and supplement them as needed to ensure that you are not limiting your growth potential (see Figure 5.1).

Goals:	To convince everyone that accidents CAN happen to them
	To set realistic expectations about how to get to excellence
	To teach the culture how to focus on the right issues and to prioritize them
	To make sure that the desired actions are positively reinforced
	To improve the model and media of safety communications
	To develop a strategy for motivating the journey to safety excellence
Methods:	Training
	Workshop
	Designing and delivering a marketing campaign
STEPS:	5.1 Passion
	5.2 Focus
	5.3 Expectations
	5.4 Proactive Accountability

STEPS to Safety Culture Excellence^{SM}, First Edition. Terry L. Mathis and Shawn M. Galloway.
© 2013 John Wiley & Sons, Inc. Published 2013 by John Wiley & Sons, Inc.

Figure 5.1   The chemistry of Safety Culture Excellence.

5.5 Reinforcement

5.6 Vulnerability

5.7 Communication

5.8 Measurement

5.9 Trust (the Bonding Agent)

STEP 5.1   PASSION

This element is what lets you know whether or not you have won the hearts of the workers to the cause of safety excellence. If they truly care and go above and beyond, it will be a matter of passion and not simply compliance. Passion comes not only from involving the heart but also from a history of making a difference. Passion is a form of motivation that involves purpose and meaning. It will be hard to develop passion without a clear strategy and a widespread knowledge of that strategy and the rationale behind it. Do not forget that passion comes from "why," not just "how."

We have discussed throughout this text the importance of getting hearts and minds involved in safety. It is not only individuals but also the whole organization that needs to have a passion for safety excellence. Passion is contagious, and your organization needs to catch it. We tend to invest our time where our passion is. We dedicate our resources and our efforts where our passion is. We will slight our interests and our hobbies, but we relentlessly pursue our passions.

Passion, like culture, cannot be directed or dictated. The best starting place from which organizational passion for safety can grow is simply having leaders sincerely state that safety is about people and caring, not just about numbers and looking good. If you have a history of pushing numbers, the perceptions will not change quickly or easily. Leaders will have to stay on-message and demonstrate their sincerity. Slipping back into the old tone or message can set back the effort significantly.

The leaders' reaction to future accidents will be another test. The official postaccident communications and the sharing of the accident investigation reports must be tuned into the human element and have a caring tone. Even if there is fault found in the report, there must also be concern for the individual, the family, and fellow workers. All safety communications must include some passionate editorial commentary including the discussions that may occur in workgroup meetings. This means that managers and supervisors must echo the message and tone of the leaders. It is easy for passion to be lost in the transfer of messages through the ranks. This means that everyone needs to be on board and reminded of the importance of passion in the communications about safety.

A potential aid in reminding and reinforcing this message can be the vision statement that comes from the safety strategy workshop. If the official vision statement for the organization reinforces the need for passion and is publicly posted, it can help keep the message in front of people. If it gets into the heads and hearts of the population, it can become a mantra that is repeated by memory and become an even more powerful reinforcement for the message.

A technique that has helped create passion at several of our client sites is translating the normal accident statistics into human terms. Instead of just saying that we have gotten our accident rates down from 3.2 to 0.9, follow that message with, "That means that before, if you worked here for 30 years you would probably expect to have a serious injury during your career. Now, only about one in three career workers will get seriously hurt. That's a lot better, but it is not good enough! If we continue that trend we can get the site to where almost no one gets hurt. That is the only goal that matters!" Such translation of numbers into human terms can turn the cold, hard facts into warm, soft emotions that build passion for safety.

Another technique that has helped build passion for safety is turning accident reports into human dramas. Instead of just describing what happened at work and the resulting injury, some organizations describe the person, their family, their hobbies and interests, and how the accident might impact these. Such translation of workplace injuries into lifestyle terms not only helps other workers realize the negative impact of accidents but also demonstrates that the organization cares about the injured worker as a person and not just a cog in the organizational gears. Not all organizations allow the name of the injured worker to be publicized, but some who do actually name the worker and show a photo of them and their family to emphasize the caring.

Although passion begins with communication, it must progress to action. How the leaders react to accidents in terms of taking care of victims and their families will either reinforce their message or negate it. Hospital visits, home visits, and making sure disability pay flows smoothly will eventually say more than words. The more the tone of communication matches the tone of action, the more the passion grows into the organization.

Case Study: I was doing an audit visit at a government facility we had been working with for several years. When we first visited the site, accident reports were shared with the workers like court dockets where injured workers were always found guilty and other workers were warned not to follow in their footsteps. On this visit, I sat in on a workgroup safety meeting and listened to a supervisor (whom we had trained) telling his team about a recent accident. He apologized that he was not allowed to use the injured worker's name, and then he proceeded to describe the man in very personal terms. He gave his age and told about his family and a recent proud accomplishment of his oldest son. He described the accident and speculated how it might impact the man's chances of returning to his old job assignment and how it might impact his life away from work. He ended his account with an optimistic report from the attending doctor and injected a little humor. The injury was a hand injury, and he quipped that luckily the gentleman had no ambitions to be a concert pianist and even more luckily he was not on the site golf team. The difference in the way accidents were reported and the way the information was received by the workers in the team was totally changed from the practices a few years ago. The human caring and passion for safety was measurably increased as well and that passion was reflected in the improved lagging indicators of safety at the site.

Application of STEP 5.1: Set the expectation that all safety communication will be personalized and humanized. Discuss techniques with everyone who communicates safety and discusses accident reports. Even urge managers and supervisors to correct workers who talk about safety or accidents in terms of blame or statistics by saying, ". . . that is true, but remember it is not just about the blame or the numbers! We have to remember how this impacts the people we care about!" Have the Safety Excellence Team (SET) review all published information about safety and offer suggestions to keep it humanized and promote passion for safety. Ask leaders to check their reactions to be sure that they reflect caring for people as well as business.

STEP 5.2   FOCUS

One of the most powerful tools in developing an excellent safety culture is called "focus." The reason for failures in many types of endeavors is either the lack of focus or focusing on the wrong things. The lack of focus can manifest itself as either not enough or as too much. Organizations that do not have a mission or vision or specific goals often lack focus. Organizations that try to work on everything at once also lack focus. The art of focusing on the right amount of the right issues can lead to success in many endeavors.

In the STEPS process, you will be asked to focus on developing the capabilities within your culture to ultimately attain excellent performance in safety. This will be a shifting focus as you take STEPS and move on to others. Eventually, you will have developed the capabilities and will be focusing (or Targeting as we will label it) your significant safety challenges and addressing them. The ability to focus the whole organization toward a specific goal or issue is a key to success. Aligned effort is a powerful force for improvement. Getting everyone on the same page working on the same things is the ultimate capability of a culture.

You begin to develop this ability when you formulate definitions for safety, accident, and other key terminology for safety. Aligning thinking must precede aligning effort. Once everyone thinks of the issues the same way and uses the same terminology to mean the same things, collaborative problem solving is enabled. Problem-solving ability is partially intuitive but can be developed to much higher levels through experience and advanced methodology. Remember, the ultimate goal is not what the culture is, but what the culture can do.

Focus is the center of expectations. Focus tells everyone what to aim their expectations toward. As the focus changes when new capabilities are developed, the level of expectations will grow to include utilizing these new capabilities to continuously improve safety. A culture with the ability to focus and expand expectations is capable of excellent safety performance. The performance may take time to perfect even after the capabilities are developed and put to use. Safety Culture Excellence is a journey and not just a destination.

After focusing on each of the key conditions and capabilities, the culture will focus on the safety issues or Targets for improvement. The main tool for selecting and prioritizing these safety targets is called Pareto analysis. This is a method of analyzing which safety solutions will impact the largest number of accidents. The ability to focus will be utilized fully when the organization is asked to focus on a particular solution to a type of accident and to help the culture to eliminate or effectively manage the risk(s) involved.

Learning to perform Pareto analysis will be an important capability to develop. It is not critical that everyone in the safety culture be an accomplished practitioner of Pareto analysis, but it helps for everyone to understand the basic premise. The idea is to identify the actions that will have the greatest impact on improving safety and to prioritize them in the best order in which they should be addressed. Working on the highest impact items makes the effort effective. Getting the maximum results from the minimal amount of work makes the effort efficient. The right use of Pareto analysis gives the safety culture quick wins, which, in turn, motivate the people involved. This has the potential to dramatically increase the amount of enthusiasm for improvement and the pace at which improvement is made.

Case Study: We were asked to help improve safety with the mobile maintenance group of a major utility company. We interviewed the safety professional in charge of the group and asked what he was focused on to improve safety. He replied that he had just received a new requirement to get all his workers in steel-toed boots so he was handing out vouchers and trying to schedule visits to the trucks and stores where they could get the right footwear. He was also responding to some complaints from facilities that the crews were not doing a good job of cleaning up and policing the areas after they completed maintenance projects. Like many safety professionals, he was reacting to the most recent and pressing issues.

We conducted a Pareto analysis of the accidents from the group for the past 3 years. Even though the workers in the maintenance group were among the best and most experienced in the company, their recordable rate was significantly higher than that of any other power plants. We discovered that even if they were perfect at wearing steel-toed boots and doing good postjob housekeeping, they would not have impacted a significant number of their accidents. They were putting their efforts into the wrong activities. After a deep analysis, we asked them to shift their focus to four items that could have impacted over 60% of their accidents analyzed. In the following year, they had 75% fewer recordable accidents, and in the next 2 years, they had none. They did not work harder on safety; they worked smarter. Focusing on the issues that have the most potential impact has a twofold impact on organizations: (1) it reduces accidents now, and (2) it creates a culture capable of developing focus that will provide sustainability in the future.

Application of STEP 5.2: The real measure of focus is not the message being broadcast but the message being received. If you interviewed a representative group of employees, how many of them could tell you what the current safety focus is? As you progress through the STEPS, are you bringing the focus of everyone in the organization with you? Some organizations formally measure this through group interviews. Others get an informal "feel" by having supervisors ask in meetings, "What is our safety focus?" and get an approximation of how many workers can respond correctly. Any such actions can potentially measure the degree of focus while also reinforcing the level of expectations that everyone know the focus and help to achieve the current goal.

STEP 5.3   EXPECTATIONS

For most workers, the expectations for safety results are clearer than the expectations for safety performance. These two are not the same thing! One of the greatest challenges of safety excellence is changing expectations from, "It probably would not happen," to, "We will take the precautions to minimize accident probability." As long as the definition of safety is "not having accidents," the perception will exist that if no one is getting injured, we are doing things right. The proper realm for expectations is not in the results but in the actions that control the results. Workers should be thinking, "Did I take the precaution or not?" and not, "Did I get injured or not?"

The concept of safe versus lucky can be helpful in establishing these expectations. The low-probability risks involved in the majority of industrial accidents only result in injuries one time in hundreds or thousands. Taking the risk tends to be self-reinforcing if you only look at the results. If you can change the mindset from the result to the action, you get workers thinking a step ahead of the potential accident. The expectation is not to be injury-free, but to be risk-free; not to be lucky, but to be proactively safe.

Acceptance of accidents or the unrealistic expectations that they will suddenly disappear can damage this goal. The necessary mix is a balance of realizing that the culture is not yet perfect and insisting on learning from accidents so that it is possible to take steps to not repeat them. Expectations must be both realistic and constantly reinforced. Unrealistic expectations lose credibility quickly, as do uncommunicated expectations. When expectations are not constantly mentioned and discussed, workers tend to assume that the expectations have changed or ceased to be important.

Expectations of Managers

It is the job of managers to set expectations for the organization. This should include expectations for safety. If managers accept accidents as inevitable, workers will tend to accept them as well. On the other hand, if managers set unrealistic expectations for instant or perfect results in safety, workers will tend to discount these expectations as nonattainable. Managers and other leaders in the organization need to realize the importance of balancing their level of expectations for safety to encourage the movement toward safety cultural excellence. The first expectation should simply be that everyone will participate in STEPS and begin to acquire the needed capabilities.

Workers look to managers for safety expectations at two critical times: the first is when organizational priorities and/or values are set; the second is when managers react to an accident. If managers neglect safety when setting priorities, workers tend to put safety at the back of the priority list. This is not to say that workers forget about safety or deliberately put themselves at risk. But safety does not stay at the forefront of their discussions, and other priorities dominate both communication and conscious thought about the job.

When an accident occurs, workers look to management to see how they will react. If the reaction is one of acceptance or indifference, the expectations for safety excellence suffer. If managers are quick to fix the blame rather than to fix the problem, the expectations tend to change to strategies to avoid blame or avoid reporting. If, on the other hand, managers react using the accident as a learning experience and genuinely take action to avoid repeating it, the expectations tend to support continuous improvement toward safety excellence.

Expectations of Supervisors

Since supervisors have the closest contact to workers among the management ranks, their expectations tend to have a significant impact on the safety culture. Workers learn quickly the supervisor's tolerance for risks and often pattern their own behaviors in reaction. Supervisors who are meticulous about safety tend to create workers who mirror that priority. Supervisors who tolerate or even encourage any level of risks in the name of productivity or competitive advantage set a completely different group of expectations. (Note: Managers who promote risk-tolerant or risk-promoting supervisors also set the expectations that such behavior will be rewarded. Since supervisors are often promoted up through the ranks, workers view the supervisor as the role model for how to get promoted. Risk tolerance among supervisors often gets perpetuated in this way.)

Supervisors set expectations in other ways that can impact safety expectations. For example, some supervisors expect excellent performance on the job and coach workers toward constant improvement. Workers who are exposed to expectations for excellence in one aspect of work tend to feel that the same expectations apply to other priorities, such as safety. Supervisors who do not set high expectations for work performance tend to also create low expectations for safety performance. It is important to remember that a safety culture is simply an aspect of the work culture. The expectations and interactions of workers and supervisors on almost any work priority will also impact safety.

Expectations of Fellow Workers

The expectations set among workers are the strongest indicator of the safety culture. The level of risk that a worker will tolerate before intervening with a fellow worker sets the "shop-floor" level of safety expectations. What is the "way we do things around here" or the expectations we have of ourselves and our fellow workers in safety? In work settings where the number of supervisors has been reduced, this influence is of increasing importance.

It is important to note that the level of expectation for safety is not the only factor that influences a worker's willingness to approach a fellow worker. There are often barriers to such communication that must be addressed before the culture is willing to communicate openly about safety issues. Even if there are not communication barriers, some workers have never had a good model of how to communicate, and it has not been the norm of the culture. This will be discussed further in Step 5.7 Communication.

The level of expectations set by the managers and supervisors can impact the level of expectations among workers over time. The most critical element of this impact is the frequency and regularity with which the expectations are communicated and reinforced. Irregular or infrequent mentions of safety expectations tend to weaken or destroy their impact. The key to expectations is to "expect" them. If you fail to expect regularly, you fail to set expectations, and workers expect that you are not serious about them.

Self-Expectations

Some have described a safety culture as what workers do when you are not looking. This can refer to a group or an individual. What do people expect of themselves in safety? How safely could individual workers perform in a vacuum without the influence of leaders and peers? It is critical that expectations for safety apply at all levels from the individual to the whole culture if excellent performance is going to result. This means that training and coaching in safety should be directed at the individual as well as the team and that workers become independent as well as interdependent in their safety performance. The tendency to see if someone is looking before taking a risk must become unacceptable. Workers must become safe for themselves as well as for the team. This is even more critical where supervision is minimal or where workers work in isolation.

One way to control self-expectations of a safety culture is to hire people whose expectations match the organization's goals for excellence. Even in a workforce of several hundred individuals, one person with a high tolerance for risks can have a negative impact on both the culture and the performance. For more ideas on how to hire employees with good self-expectations in safety, see the article *Hiring for*

Safety: Risk Takers Need Not Apply from Industry Week April 2010 (or from the ProAct Safety^{SM}* web site, http://www.proactsafety.com)

Additional Considerations: Hiring For Safety Excellence   Since the beginning of the Industrial Revolution, organizations have continued to enhance their preemployment practices. Recent additions have included background checks (personal, professional, criminal, and credit), reviewing driving history, assessments of competency (demonstrable knowledge, behaviors, and skills required to perform specific tasks), and personality profiling (tests to determine character, patterns of behavior, thoughts, and attitudes).

A number of businesses now utilize advanced strategies to determine a candidate's thoughts on preventability and overall attitude toward safety, such as the locus-of-control scale conceptualized by Julian B. Rotter in 1954. This questionnaire attempts to determine an individual's perception of the control over events that surround them by presenting statements indicating whether they are either more internal (I am in charge of my life. I control my destiny. I can make a difference.) or external (What will be, will be. If it is to happen, it will. It is up to others.).

Other businesses have expanded or supplemented preemployment approaches by determining a potential employee's degree of self-efficacy, a concept developed by Albert Bandura. Self-efficacy is most commonly defined as a person's belief in their own ability to achieve a successful outcome. While these are excellent tools and have offered significant insight into the psyche of the potential candidate, the approaches can sometimes be cost-, time-, or complexity prohibitive.

Individuals who maintain responsibility for site succession planning and employment require simple strategies to ensure that the right safety-minded candidates are selected, especially when the aforementioned tools are unavailable. Regardless of industry, resources have become scarce, and the standard adage of "do more with less" has become more pervasive.

There is no perfect list (quantity or quality) of questions that gather the complete, vital insight into the safety attitude of a potential candidate. Furthermore, providing the ideal predefined questions is counterproductive to the purpose of this book. Every company will be on a different point of their journey toward operational or cultural safety excellence. In addition, many organizations and certain countries have legal guidelines for what may or may not be asked during an interview. Thus, the nature of the questions could be different. The questions will also differ based on the level and responsibilities of the position and applicant.

The following four sample questions are posed to demonstrate the possible value that open-ended, safety culture interview questions could bring to your company during the hiring process.

Question 1: How Would You Define Safety?   Does the individual define safety with the inaccurate and unfortunate standard statements, "not getting hurt" or "going home the way you came in"? Or do they have a more proactive mindset that indicates thought-provoking preventative strategies? Do they respond with positive statements, such as "controlling and eliminating risk"?

* ProAct Safety is a registered service mark, U.S. Registration No. 4,011,455.

Question 2: What Role Does Safety Play in an Organization? Unfortunately, some people feel that safety is just another thing to do, rather than the way to perform work. Look for answers that identify safety as a guiding value that compliments operational activities. Responses that indicate an understanding of how safety adds benefit to the employee's personal lives, their families, the site morale, community standing, customer perception, and overall market position would be desirable.

Question 3: What Does It Take to Reach and Sustain Zero Injuries? During this answer, look for positive indications of a behaviorally demonstrated belief in the journey (i.e., body language, facial expressions, and word usage). Those who are passionate about safety being part of their job will likely have given thought to this question prior to it being asked. If a large percentage of people within the culture do not believe in the destination, it will be extremely difficult to get there. Moreover, if the leaders or potential new leaders do not feel that it is possible, the difficulty increases exponentially.

Question 4: What Do You Think the Safety Roles, Responsibilities, and Expectations Are for Someone in This Position? Answers for this can range from "making sure people do not get hurt" to behaviorally descriptive strategies that help create the desired outcome of zero injuries or 100% safe. Preferably, the candidate would respond in a way that outlines the actions an individual in that role would need to perform on a frequent basis to create the ideal environment. The candidate could even outline how the expectations could be measured.

Questions such as these have assisted multiple hiring authorities in identifying desirable and undesirable attitudes, perceptions, and personal safety perspectives. Determining this prior to new-hire onboarding practices helps an organization answer a critical, yet often overlooked, question: "Am I hiring a person that complements our evolving safety culture, or am introducing new risk into our company?" You would not know until you ask. If safety is truly an integral value within your organization, do not wait to demonstrate it during orientation. Incorporate the communication of safety values into your hiring practices. Consider the message this sends to prospective employees: Safety starts before day one, so risk takers need not apply.

Expectations of the SET

This is the group leading the journey toward safety cultural excellence, so their expectations and their examples will be of crucial importance to the process. However, if they are the only ones on this journey, the destination will not be reached. It is important for the population, at all levels, to sense the same expectations and hear the same message uniformly from the process leaders, the formal leaders, and each other. The expectations should send the message, "We are starting on a journey to excellence. It will pass through destinations in which we will develop conditions and capabilities conducive to excellence and end using those capabilities to identify, prioritize, and address our safety challenges. We will be patient but relentless in our journey toward better and better safety performance and never lose the goal of the eventual level of excellence we desire to achieve."

Case Study: In 2009, on a cold, windy winter evening at a client facility, a long-term employee dressed in dark clothing, covering his head with a hood, had just entered the roadway after walking between two buildings. He was, unfortunately, struck by a reversing delivery truck, and this resulted in his death. The facility had just completed a change in traffic patterns, rerouting the incoming flow from entering from the left, to reversing from the right. It was believed that the employee did not hear the reversing alarm coming from the vehicle. Moreover, it is likely that the individual habitually looked left where the traffic previously came from.

While certainly not the root cause, the organization believed that reflective vests would have aided in the driver seeing the pedestrian. As a result, the organization immediately decided to institute a gate-to-gate policy of reflective vests at all of their facilities. Six months after the fatality, we were asked to conduct cultural assessments at 11 of their locations. While reflective vest compliance was not on their immediate radar of focus, they did quickly notice a very low level of compliance and support for the new rule throughout the company.

While visiting a location in Ireland, however, they quickly noticed 100% compliance with the rule. During group interviews, they indicated a strong support for the policy. One lady stated (in a wonderfully thick Irish accent), "Wearing the vests is just a thing we do. In the mornings, after brushing my teeth, I put on my vest and head to work. I guess it has become a habit." We were insatiably curious to determine why this site had been able to accomplish what the other 10 could not. The answer soon became clear: effective communication.

The safety manager at this location was close friends with the safety manager at the site where the fatality occurred and had personally known the individual whom had lost their life. When the Irish safety manager was instructed to implement the new rule, he met with employees in small groups across all crews in the organization. He showed the groups a picture of the employee who had lost his life and his family members. He then stressed how he knows many of his employees' families and never wants to be in the position of his friend, having to tell a family that their father would no longer be coming home. He then spoke about how the family was now struggling, following the loss of their father.

The Irish safety manager took the time out of his already overwhelming work load to emphasize in a compassionate tone that the organization truly believes that a reflective vest could have been enough of a prevention tactic to have stopped the chain of events from turning out the way they did. The reason for acceptance for change on this rule was clear, and he made the reason for the change personal.

Communicating effectively is difficult; as a result, a common perception that tends to exist is that employees feel that most communication to the workforce for safety-related initiatives is general or basic, or even nonexistent. It is to be expected that this results in employees not supporting, understanding, or having any sense of ownership in approaches to improve safety or other areas of operational excellence.

Excellence does not happen through blind compliance. We *do* want our employees to think and question things they do not understand. For this, they need access to information. While it is not always possible or practical for employees to be involved in all changes, it is critical for support that they understand, at minimum, the rationale.

Application of STEP 5.3: The basic tools for setting expectations are communication and follow-up. It is critical that communication be consistent and continuous. Everyone must stay on message throughout the process if the expectation is going to be received and believed. It is also necessary that follow-up occurs after each STEP to ensure that everyone is on the journey and keeping pace with the organization. If the message falters or the follow-up fails to happen, the level of expectation either reduces or disappears. In this world of constant change and shifting priorities, many people are sensitive to the "message of the day" and know that what was important yesterday might not last. Safety excellence is a goal worth pursuing and that pursuit must stay the course if it is going to be successful.

The message and follow-up needed to set the expectations should come from every level of the organization. But it should also come from every media used for communication. That means that the STEP in progress should appear in newsletters, on bulletin boards, on electronic media boards, be mentioned in meetings, and be handed out as flyers or any and all other media in use at the site. Some sites actually develop and use new media for the Safety Culture Excellence initiative. Cooperation from corporate level newsletters and media can also be useful, not only for communication but also for recognition and motivation for the site(s) and people involved.

STEP 5.4 PROACTIVE ACCOUNTABILITY

If your culture will tolerate risks as long as they do not result in injuries, that must be changed. Accountability must move upstream, and workers must be held accountable for their actions even when they do not result in injury. The tolerance for risk taking must decrease, and this must start from formal accountability systems and move to the informal interactions among workers. You cannot manage undesired results except by managing the behaviors that either cause or prevent them.

If you have not already implemented safety roles, responsibilities, and results (SRRRs) as we suggested in the formation of your safety strategy, now is the time to begin to do so. Excellence in safety demands that every person in the organization know exactly what they should be, do, and accomplish in terms of personal and organizational safety performance. The expectations from STEP 5.3 can be a starting place to develop SRRRs and can supplement any job descriptions you may already have. Remember, these are not overall job descriptions; they are "safety" job descriptions.

The greatest challenge for some organizations is moving away from strictly reactive accountability. Historically, accidents and/or near misses have been what triggered the need for accountability in many organizations, and they may need a new trigger to hold people accountable for actions that do not result in accidents. The SRRRs can be a part of that trigger. If you hold annual or more frequent job evaluations, they can now become a part of proactive safety accountability. But these are seldom sufficient alone. Everyday, accountability needs to focus on safety actions that can prevent accidents. Both the organization and individuals must overcome the reactive definition of safety and the reactive accountability it fostered. Safety must be defined as what we do to prevent accidents not whether we have an accident. Accountability needs to make the same move from results to actions.

This shift in accountability must be a unified message followed by unilateral action within the organization. Leaders should send the message. Managers should echo the message in their meetings and through their direct reports. Supervisors should send the message often in meetings and in their everyday coaching in the workplace. Workers should help each other stay focused and take precautions. When official performance appraisals happen, they should contain the same message and expectations. Every communication and every action taken should set the same expectations and hold everyone accountable for making their contribution to the effort.

Case Study: A company asked for our advice immediately after they received Occupational Safety and Health Administration (OSHA) citations for lockout tagout (LOTO) violations. They had trained workers in proper LOTO procedures and could produce the signatures of the workers involved on the training sign-in sheets. They also showed that they had punished a worker who was injured for a LOTO violation. But the OSHA inspector questioned supervisors and workers on the floor and found that there was no accountability or expectations on a regular basis for following LOTO procedures, only training and reactive accountability. If the supervisor had proactively held workers accountable for following LOTO, there would have been no citations. We found numerous other common practices that could produce both accidents and violations that were not being addressed except after accidents. We recommended a robust proactive accountability system that was adopted by the organization with excellent results.

Application of STEP 5.4: Many organizations already have taken action to promote proactive accountability for safety and will only need to look for ways to improve their efforts. However, if you have not done anything yet to address this issue, the best approach is a strong directive to everyone that this is the new standard of performance. Trying to make this happen slowly or by level has not been successful for most of our client organizations. Sending out a directive and following it up with meetings and official communication have proven better for most organizations. Developing SRRRs can take time and may need to be an ongoing project. These are living documents and often undergo yearly reviews and changes anyway, so starting them as basic documents and letting them grow usually do work well. If you are depending on the SRRRs to give substance to your proactive accountability effort, you should consider the timing of this STEP to correspond to the SRRR development project.

STEP 5.5   REINFORCEMENT

Excellent organizations are keenly aware of what behaviors are being reinforced, even if the reinforcement is not intentional. Sometimes recognizing reinforcers is as easy as asking the workers. Sometimes the reinforcement does not become obvious until it manifests itself in accident reports. Leaders need to start asking, "If we do this, what risks might it reinforce?" and "What can we do to reinforce the taking of key precautions?"

Behavioral scientists have pointed out that behaviors that get reinforced get repeated more often than those that are not reinforced. Yet many organizations do not realize that making a policy or statement does not necessarily constitute reinforcement. In fact, many organizations inadvertently reinforce many behaviors they do not intend to nor desire to have repeated. Many safety incentive programs, for example, actually rewarded (and thus reinforced) not reporting accidents. Workers received a reward for not having an accident, and therefore they did not report the accident to get the reward.

It is important to develop a deep understanding of what reinforcement really is and how it works. Behavioral scientists have built a little "fail safe" into their terminology by the way in which they define the term "reinforcement." They say that anything that increases the probability that a behavior will repeat is, by definition, reinforcement. You cannot say that reinforcement does not work without the behavioral scientist responding that; if it did not work, it was not reinforcement. In short, the whole criterion for whether or not something is reinforcement is the change in probability that the behavior will repeat. Reinforcement follows a behavior in chronological order, but the knowledge of past consequences and anticipation of future consequences can actually precede and prompt behavior.

Skinner and other founders of behaviorism and classical and operant conditioning defined a number of standards for behaviors such as the rate, duration, magnitude, latency, and so on, but for the purposes of building an excellent safety culture, let us stay on the simple plane that reinforcement makes someone more likely to repeat an action. There are many different classifications of reinforcement but let us concentrate first on two main types: positive and negative reinforcement. There is a common mistaken idea that negative reinforcement is punishment. Not so! Remember the definition of reinforcement: something that increases the probability that the behavior will repeat. Punishment reduces the probability that a behavior will repeat and is not therefore a form of reinforcement. The terms positive and negative generally refer to whether the reinforcement is accomplished by adding or subtracting. Giving a reward for a behavior is positive (adding) reinforcement. "If you will lead the safety meeting, I will buy you lunch." Taking away extra duty for a behavior is negative (subtracting) reinforcement. "If you will lead the safety meeting, I will get you off the cleanup detail."

Not all reinforcement happens directly or during interpersonal communication or interaction. The very way a job or task is designed, or an incentive system, or a style of management can be reinforcing or not so. A task that is properly defined and works well can reinforce workers to do it a certain way, which makes the task easy and trouble free. Work procedures that require awkward actions or take more time to finish the job can be punishing and actually reinforce noncompliance. An incentive program can make it desirable to do certain things to reap the rewards. Managers and supervisors who praise good performance can actually reinforce worker behaviors. Such reinforcement can make a workplace pleasant as well as safe or effective if it reinforces the kinds of behaviors that create these states. However, if the consequences in play in the workplace punish or make it difficult to do these behaviors, the opposite can happen.

The coaching model of communication and performance management described in the Climate section, Milestone 4, is one of the strongest ways we have found to establish effective positive reinforcement at a personal level. Doing so tends to start and encourage it to happen at the organizational level. Giving positive reinforcement also heightens the awareness of how other factors outside personal interaction tends to reinforce or not. As the organization becomes more and more aware of reinforcing factors, it can better align them to reinforce the desired performance and not reward unwanted behaviors.

Case Study: We were performing an assessment of an organization's safety culture. We were going through our list of questions with our first group of workers being interviewed and asked, "What would happen if you took a lot of risks but did not get injured?" The unanimous answer of the group was, "You would get promoted to supervisor!" We found throughout our assessment that the organization rewarded what they called "safety," which was defined as "not having an accident" for a significant period of time. In the most common job in the organization, there were some safety shortcuts that involved relatively low-probability risks that could save time and increase productivity. Those that took these shortcuts and lucked out appeared to managers to be both the safest and the most productive workers. They regularly got promoted to supervisory positions. The organization was not deliberately reinforcing safety shortcuts, but that was the result of their actions. We have found many such instances of good intentions inadvertently reinforcing bad safety performance.

Application of STEP 5.5: One of the most systematic ways to discover what organizational issues reinforce what kind of safety performance is through safety coaching or behavioral observations. When a worker is observed doing something safely or in a way that may be at risk, if the coaching supervisor or fellow-worker observer sees the behavior and asks why it is happening, the answers to these questions can reveal what reinforces workplace behaviors. When leaders and managers make decisions about workplace or workflow design issues or procedures, they often cannot fully predict how they will impact safety performance. Reinforcement can also be detected in perception surveys and focus-group interviews or through more formal studies. The critical issue is to change practices when it is discovered that they reinforce the wrong kind of performance and constantly look for more and better ways to reinforce good safety performance.

STEP 5.6   VULNERABILITY

One of the greatest enemies to safety excellence is the perception that accidents "do not happen to me" or that they "always happen to the other person." The most likely type of accident to happen in many organizations is the kind that no one thinks will happen. The feeling of invulnerability often results in actions that tempt fate. Workers who do things that have gotten others injured and "get away with it" tend to start a pattern of thinking that they are immune from the dangers that plague others. They tend to test this theory with occasional actions that take them to the edge of danger. If they escape injury, these tests get more frequent and comfortable. Over time, they can actually become habitual and not require conscious thought at all.

Many researchers have listed the issues and perceptions that tend to encourage workers to take risks. Such issues as familiarity with the tasks, fatigue, monotony, trusting the judgment of fellow workers, and supervisors are often mentioned. Perceptions such as, "this is not likely to happen," or, "I just need to get the job done quickly" can also play a part in a person's willingness to take risks. However, the one commonality of all these issues and perceptions is that they involve risks that have a relatively low probability of resulting in an injury. These low-probability risks are involved in most industrial accidents at sites with active safety programs and are at the root of perceptions of invulnerability.

Workers tend to learn the details of safety from experience. Taking low-probability risks regularly tends to feel comfortable until the probability plays out and an injury or near miss occurs. The problem is that low-probability risks and safe actions both result in no injury most of the time. Experience does not often distinguish the difference until the injury event, which is too late.

For this reason, it is crucial to instill a sense of constant vulnerability into the workforce. Workers who feel vulnerable take safety seriously and regularly take precautions to increase their safety zones. Workers who feel safe feel free to take actions that decrease their safety zones because their experience tells them that they can do so successfully. Consequences reinforce behaviors, and low-probability risks usually have satisfactory or even desirable consequences most of the time.

Case Study: Manufacturing Company—We were asked to conduct an audit of the safety processes in place at a division of manufacturing sites for a large company. Each of these sites had more than one manufacturing facility in operation. At each location, one of the facilities manufactured a hazardous product, and the other facility manufactured a fairly safe product. In every instance, the facility that manufactured the hazardous product had significantly better safety performance than the other, seemingly safer facility. As we conducted the interview portion of the audits, it became obvious that the workers at the facilities that manufactured hazardous products had a much higher sense of vulnerability than the others. They came to work everyday knowing that their actions could result in catastrophic results, both for themselves and for their fellow workers if they did not strictly follow their safety protocols. The workers at the other facilities felt no such compelling reason to be meticulous at safety since their product was not viewed as particularly hazardous, and the results of this showed in their safety performance results and costs. Both groups worked for the same company but produced different safety performance; the difference was the sense of vulnerability.

Application of STEP 5.6: Some STEPS that can help to instill the sense of vulnerability in your culture include the following:

- *Adopt and communicate a "definition of safety."* The issue of vulnerability can also be described as distinguishing between "safe" and "lucky." This is where the definition of safety discussed earlier becomes evident. If workers define safety as "not getting hurt," they continue to mix the notions of safety and luck. Either one can "not get you hurt" most of the time. Excellent safety cultures realize the potential of low-probability risks to result in accidental injury and regularly take precautions to prevent them, even though experience tends to minimize the danger.

- *Communicate accident data cumulatively.* When accidents and near misses happen, communicate them to the workers who can benefit from the information. Make sure to tally the information not only by accident type but also by which precaution(s) could have prevented them. Keep workers constantly aware of which risks most often result in safety issues and which precautions can keep them safest from these issues.

- *Make safety personal versus abstract.* When safety communication focuses on statistics (i.e., recordable rates and severity rates), workers seldom get emotionally involved and often fail to understand the importance of their efforts to themselves and others. Talking about individuals and how accidents impact them, their families, and their lives is more likely to evoke emotional involvement and elicit discretional effort from employees to improve safety. Stories and testimonials from accident victims and their family members can demonstrate the fallacy of invulnerability. Almost everyone involved in an accident had the perception that it would not happen to them, and hearing about the personal shattering of that perception can be both touching and insightful.

STEP 5.7   COMMUNICATION

All the other elements of Safety Culture Excellence are either enhanced or weakened by the quality and quantity of communication within the organization. Keeping people knowledgeable and focused can only happen when the important information is both sent and received. Almost every organization should address this issue in the strategy and also take the STEP of reassessing and looking for ways to further improve communication in safety. Almost every worker we have ever interviewed perceives the priority of safety in the organization from the volume of communication about safety versus other priorities. Leaders who regularly and effectively communicate about safety tend to make it improve.

If a culture is going to become excellent at safety (or almost any other aspect of work), it must be a communicative group, and communication must be effective and frequent. If your organization lacks media or systems for communication or does not have a history of effective communication, you may have some groundwork to do before you proceed with the techniques described here. Even if you are already reasonably effective at communication, you might want to explore new media and techniques that are coming into more common use.

Communication is the way a message gets from one person or group to another. There are always senders of the message and receivers of the message, both of which have to be turned on and tuned in at the same time. Just like a radio or television signal is transmitted by the station and received by the radios and televisions. If either is not turned on or tuned in, communication is not accomplished. Many people believe that communication is simply the act of sending a clear message. But sending is incomplete if receiving does not happen. Have you ever talked to someone who did not seem to be turned on or tuned in? Has anyone ever tried to talk to you when you did not want to be turned on or tuned in? Have you ever turned off the radio or television when you no longer wanted to hear or see the message?

Before we jump into the details of how to effectively communicate, let us discuss basically what communication is and what makes it effective. First, more than just information is transferred in good communication. Good communication is rich with meaning and context and emotion, not merely data. Think about the great speeches you have heard from leaders of government, religion, business, and science. You tend to come away from these with much more than information. Why? It is because they reached out to more than simply the storage capacity part of your mind. They reached out to your imagination, your feelings, your values, and your desires. This does not mean that every transmission of data must be a sermon or lecture or motivational message. It simply means that we need to communicate rationale as well as command, context as well as message, meaning as well as data. The old saying to "speak from the heart" approaches this thought from one aspect. Really "meaning what you say" approaches it from another.

Why is the richness of communication so important? Communication expert, Albert Mehrabian, in his book, *Silent Messages*, stated that the truest measure of communication is "believability." In a landmark study, he discusses the three elements of spoken communication. He labeled the three parts the verbal, vocal, and visual. The verbal element is the message itself, that is, the words you choose to convey the message. The vocal element is the tone quality of voice when you deliver the message (volume, pitch, speed, etc.). The visual element is what we often call body language, that is, what people see in the speaker's face and body as the message is delivered. When messages are inconsistent and thus not believable, these elements can actually contradict each other. For instance, a speaker whose words say he or she is excited to be here, but the tone and body language show no enthusiasm. In such communication, the verbal only contributes 7% to the effectiveness of communication, the vocal contributes 38%, and the visual contributes 55%. How often do speakers choose their words with great care but neglect their tone and body language?

There are four other qualities that contribute to the effectiveness of communication:

- *Objectivity.* Communication that is based on bias, speculation, or generalization is almost always less believable than communication based on the facts and free of point of view. Even when stating the facts, certain words can seem to be labeling or classifying and can hurt the objectivity of the message.
- *Clarity.* Good communication is free of dual meaning, and the elements of communication work in harmony. This means that the words are chosen carefully and that the context in which we use them, including tone and body

language, does not contradict each other or confuses the meaning. Almost every word in the English language has more than one meaning, and we glean which meaning is intended by the context in which we use the word.

- *Two-way.* Good communication travels back and forth between individuals, not always from the same one to the same one or ones. Two-way communication allows for clarification and probing of meaning. It also involves more senses than listening which makes it more interactive and more "sticky" or memorable. Two-way communication ensures that both the sender and the receiver of the message are active. Involvement in a conversation also creates more ownership of ideas discussed, and more ownership usually results in more active participation.

- *Nonthreatening.* When communication becomes threatening, the receiver becomes defensive, and the message is not received openly or completely. When we feel threatened, we trigger what psychologists call our "fight or flight" response. In conversation, this means that we either verbally fight back or simply withdraw from the conversation. This keeps it from being two-way and compromises the quality of reception of the message.

Communication Media

Most of the information about communication so far has been about spoken communication between individuals or groups, but there are many more ways to send a message.

- *Communication in meetings.* Almost all organizations have meetings, either as a whole population, or in work teams, or by shifts, or a combination. Sending out messages to be delivered in these meetings can be an effective way to tie everyone together while recognizing the differences that might exist between various teams, areas, tasks, and so on. Shift start-up meetings and/or toolbox/tailgate meetings are often timely to set the tone and focus for upcoming work. Sometimes it takes a systematic approach to maintain the consistency of the message when it is delivered by different people at different times and in different settings.

- *Written communication.* Organizational or site newsletters, flyers, bulletin board postings, digital message boards, and others can be effective as well. Some newsletters are taken or sent to workers homes and can get information to the family as well as the worker. Writing can take more time and effort than spoken language, but it can also have a longer impact as it does not disappear as quickly as spoken words.

- *Broadcast communication.* Some organizations have speaker systems, intercoms, teleconferencing, and other media for sending a message out to multiple people and even multiple sites at once. This can be less personal but can have better outreach and more consistency of message if utilized effectively.

- *Computer-based and/or Internet communication.* E-mail, computer-based training (CBT), webcasts, podcasts, blogs, social media, and other forms of communication have come into popular use for safety communication in recent years. These can be a good or bad fit for the organization depending on the access to computers and the Internet. They can be highly effective to reach lone workers separated by distance, time zones, and even language barriers.

Barriers to Communication

When designing an effective communication strategy, it is important to consider what can hamper or disable communication. In addition to physical barriers such as distance and noise, there can be systems barriers such as having people on different shifts or sending information through supervisors who differ in their communications habits and skills.

- *Bottom-up communication.* A common barrier to communication in safety is simply people who shield their boss from bad news. Often, leaders think their direct reports are sensors when they are really censors. It is important to let everyone know that the organization and its leaders value and need accurate information, even when it is not favorable. Good leaders tend to talk to people at every level in their organization rather than relying on information flowing up through the structural ladder.

- *Top-down communication.* Likewise, it is important to have information flow uniformly down the structural ladder. This is where new media can be of help. Using teleconferencing, videos, webcasts, or podcasts can get the same message to everyone delivered by the same messenger. This can potentially help to overcome the problem of message drift as well as to ensure that the message gets through. We encourage the use of multiple media for communicating official information to avoid having barriers to one media completely compromise this important flow of information.

- *Lateral communication.* Excellent safety cultures have a high degree of communication among workers. Achieving this level of communication will necessitate removing barriers, if they exist, and reinforcing the practice of "talking about safety" among each other on a regular basis. STEPS has a methodology to help accomplish this, but it will help if you identify the barriers prior to implementing. Are there physical, structural, or organizational barriers between workers? If so, can you facilitate opportunities for workers to come together and discuss safety? Is talking to each other about safety something workers already do or not? Are such discussions considered butting in to someone else's business? Do workers simply complain to each other about safety? Understanding your starting point can help to improve the quality and speed of communication improvement.

What Needs to Be Communicated

Once you understand how communication works in your organization, what media are available, and what barriers need to be overcome, you need to develop a strategy to improve the system and to deliver the right information effectively. Some of the information to consider include the following:

- *Communication of focus.* As you progress through the STEPS, you will find your organization changing your focus regularly. As you do, it becomes of increasing importance to keep everyone informed, aligned, and "on the same page" about what you are working on and how that effort is progressing. Some organizations create new media (such as handouts and meeting opening statements) or special sections of media (such as regular columns in newsletters or places on bulletin boards) to keep everyone informed as to what STEP is the current focus. Mapping the STEPS and tracking the progress through the STEPS is often an effective way to keep everyone progressing together and informed.

- *Communication of accident data.* The goal of Safety Culture Excellence is the elimination, as far as possible, of accidental injuries. The occurrence of injuries is therefore critical to tracking progress as well as to maintaining focus. Most organizations treat accidents as individual events in their communication strategies. They relate the details of the accident and seek to learn lessons to prevent similar accidents in the future. As you progress in STEPS, you will be focusing on categories of accidents or on precautions that can prevent certain types of accidents. Therefore, you need to keep all communication of accident data in the larger context of what you are trying to accomplish. It is important to communicate constantly the categories of accidents and the tally of how many accidents in total and by category.

- *Communication of progress.* Communication on Focus tells everyone which STEP you are on. Communication of progress tells everyone how far through the STEP you are and where you are in the map of total STEPS. Once you have passed the basic STEPS to help you control conditions and develop competencies, you will be focusing on specific safety issues and will want to track how many accidents you have had in the focused category as well as how the numbers are trending and responding to focused efforts. Remember that progress toward goals, when communicated effectively, is not only informative but motivational. An excellent safety culture is one that can focus and improve. Seeing visible progress is part of what motivates that effort.

Case Study: We made a proposal for a safety-improvement initiative at a public utility district (PUD) in the Northwest United States. Since it is a public entity, the leaders had to get permission from the voters to fund the project. While we were waiting for the wheels of progress to turn, we did our Pareto analysis on the accident data from the PUD. We identified four precautions that could have theoretically prevented 71% of their accidents over the past 3 years. The safety leader started a newsletter for the safety-improvement project and began to report how each month's accidents fell into the four categories we discovered. By the end of 120 days, there had been four monthly newsletters spelling out the details of how we proposed to improve safety, and there had been a 95% correlation of accidents to the focused categories. A lot of the workforce population was already educated and sold on the approach and the wisdom of focusing on the four precautions that could impact the majority of the accidents. The good communication continued after the project began, and the newsletter became the chronicle of the focus, progress, and ultimate success of the process.

Communication and Marketing   Communication is a way to convey information. It is also the primary tool of marketing. If you developed a marketing plan as part of your safety strategy, communication is the primary tool to carry out that plan. The goal of your marketing plan should not be to get your workers compliant or open or willing or participative or submissive. The goal should be to get them SOLD on creating an excellent safety culture. To accomplish this goal, consider two other tools of communication that can help.

Tool 1: Personalizing   People respond emotionally to change prior to responding logically. So why do we sell change by starting with the logic and organizational benefits? Good definitions can create clarity on the "what" and "how" of safety excellence. Good definitions do not answer the question "why"? Every organization has its own unique reasons for seeking further improvement in safety. Sharing this rationale with the workforce can be a powerful incentive to participate and to excel, especially if the why includes the "what's in it for me?" (WIIFM). You can dictate this course of action and get workers' minds into the effort, but their hearts are always volunteers. Many scientists and philosophers have pointed out that the brain tells us to think, and the heart tells us to act. If you can get your workers involved emotionally as well as mentally, you will get a different level of effort.

One way to create emotional involvement is to "personalize" safety. Charitable organizations learned decades ago that if you want people to donate to abused animals or crippled children or starving people on other continents, the best way to do so is to show photos or videos of the victims that touch the hearts of the potential donors. Talking about numbers and statistics fails to elicit the same emotional response as making it personal. We think about numbers, but we care about people and animals.

Case Study: A client organization was planning a kickoff event for their safety-excellence initiative, and they realized that making it personal could have a good impact on the event. After much discussion, they decided to purchase a pair of paper-framed glasses with red plastic lenses (similar to the old 3D glasses from cinemas) and have their safety project logo and name printed on them. They handed out a pair to each person attending the kickoff sessions and asked them to put them on. They showed a PowerPoint slide with all the facts and figures of safety: recordable rates, severity rates, lost days, costs of workers' comp, and so on. The moderator told the group, "This is the way we used to think about safety." Then they asked the attendees to remove their "rose-colored glasses" and look at the same slide. Only now, without the red lenses, all the data were gone, and you could only see a mural of all the people who worked there and their families. The moderator told the group that they were now going to think about safety as a duty to keep the people we love and work with safe. They showed a video in which several workers who had been injured and their families were interviewed and asked how the accident had changed their lives. The president of the division of the organization added his heartfelt support for the effort and wished everyone great success. What had just happened? Safety had been personalized and the hearts of many were won over to the new safety-excellence effort, which became a great success.

Tool 2: Repetition Once you have developed your message and marketing strategy, it is important to do what marketers call "keeping on message." This means that the message needs to be repeated often and consistently to reinforce and emphasize that this is important and it is not going away. Many safety efforts have relied on changing programs to keep attention and awareness on safety and have been labeled the "program or flavor of the month." Any new safety effort can fall prey to old thinking and be considered to be the latest fad and something that will run its course and disappear. Keeping your message consistent and frequent can help you avoid or overcome association with past programs or history.

Most good marketing programs utilize multiple media to distribute their message. You can deliver your message in safety meetings, before safety training sessions, at toolbox or tailgate meetings, in official communications, and so on. But challenge yourself to look for other media also. Can you use bulletin boards or information monitors? Do you have a company or site newsletter? Do you distribute pamphlets or put insert material in paychecks or benefits information? Would your community newspaper, radio, or television stations carry stories about your safety efforts as human interest or community news?

Case Study: One site had such a problem with their safety-improvement effort being referred to as the "program of the month" that they put up banners throughout the workplace on the program's third birthday that said "Our Program-of-the-Month is 3-Years Old—Happy Birthday!" They talked about the initiative in safety meetings and wrote about it in their company newsletter. They shared the trends in their safety lagging indicators before and after the initiative and pointed out that this is the only thing they are doing different in safety across the time periods. They recognized the people involved directly in the initiative and showed how the process KPIs had grown toward their targets and how that corresponded to the improved lagging indicators. They changed a lot of minds and attitudes with the effort and breathed new life into their "program-of-the-month" safety-improvement initiative.

Application of STEP 5.7: Review the material in this section either as a reading assignment and discussion or as a workshop and evaluate your safety communication (including your safety marketing strategy) against these criteria. Look for ways to communicate more frequently through more media in a more effective way. Remember that the most meaningful measurement of effective communication is the message that is received rather the message that is sent.

STEP 5.8   MEASUREMENT

There is an old saying that what can be measured can be managed and what gets measured gets managed. Safety has suffered for decades from limited measurement.

The lagging indicators are important, but they do not tell the big picture nor are they prescriptive. They tell you if you have a problem, but they do not diagnose the problem or reveal the solution very effectively. Exploring more meaningful and holistic metrics for safety is a critical initiative that every organization must undertake if they are to reach Safety Culture Excellence. As we mentioned earlier, it is not just a matter of leading and lagging indicators. It is a matter of failure and success metrics. We tend to measure what we do not want and fail to measure what we do want. Good measurements aid improvement and motivate people to want to improve. How excited are the individuals in your organization about your current measurements? Do your measurements prescribe how to get better and describe precisely why you are achieving your current results?

Measuring Your Starting Place

The first step of navigation is to precisely determine your current location. Until you know where you are, it is difficult to plot a course to your desired destination. Likewise, it is a good idea to measure where your culture currently is before plotting a course to safety excellence. The STEPS process will ask you to assess your status on each targeted area as you address it, but such evaluation will not give you the big picture of where you are in relation to the whole journey.

To get this "big picture" of your starting place, you should have either completed the assessment activities described in Milestone 2 or had a cultural assessment professionally performed. This activity will prove valuable both for directing your efforts and for measuring your progress. Setting realistic expectations in the beginning will help avoid disappointment and loss of commitment as the journey progresses. The more clearly everyone sees the current location, the desired direction, and the destination, the more support and participation you will get.

Progress Through the STEPS

The first part of the journey to Safety Culture Excellence involves completing STEPS to develop the Clarity, Climate, Chemistry, and Control needed to address safety challenges. Each STEP is described to help you determine your culture's current status and any improvements needed. As you move through these STEPS, you will find your culture more poised and equipped to handle safety challenges. Mapping your journey through these predefined STEPS can be a meaningful measurement. It can help everyone see more clearly the path the organization is taking and measure progress toward the goal.

Focused Action Plans Developed

After you have completed the predesigned STEPS, you will begin to define your own next steps that will be to address your specific safety issues and challenges. Using the methodologies prescribed, your culture will learn to identify, prioritize, and address specific safety issues. Each of these units of improvement will be an ultimate STEP toward excellence and will involve specific action plans. A metric of progress is simply how many of these action plans you have developed. Be careful not to drive quantity and sacrifice quality. Some action plans will take longer to complete than others, but you will be guided to move the quick wins to the front of the queue so that you can make progress and motivate the process. Also, your culture needs to hone its skills at solving safety problems, and the best way to accomplish that is to start with easier challenges and work your way toward the more difficult and complex issues. That way, the skill increases more proportionately with the magnitude of the challenge.

Focused Action Plans Completed

As your process progresses, you can measure the number of action plans completed. This number will become a more and more meaningful metric of the maturity of your process. The ability to successfully address safety issues is a key to Safety Culture Excellence, and the number of issues addressed is a good process indicator. As with all process indicators, the real goal is to impact results, not just to measure activity.

Measuring Perceptions

A valuable part of your overall evaluation is a beginning measurement of perceptions about the current status of safety efforts, culture, and performance. This will help you determine your starting position and will be a valuable benchmark against which you can measure future progress. Changes in performance will begin with changes in perceptions, and improved performance will further change perceptions. The measurement of perceptions is something we recommend on a periodic basis as a part of a larger safety metric. To regularly measure perceptions in a cost-effective manner, it is not possible to use most off-the-shelf perception surveys. In addition to prohibitive costs, they are not designed to measure all the elements of perception critical to developing an excellent safety culture.

Grounding Perception Measurements

One weakness of standard perception measurement lies in the fact that there are two major types of perceptions: accurate and inaccurate. The standard survey only indicates what people perceive and the degree of agreement or disagreement among those surveyed. Other data can help to determine the accuracy of perceptions. For example, what if most people surveyed perceive that safety training is effective in preventing accidents, but the majority of your accident investigations discover that workers were not adequately trained to handle the risk? This perception is not accurate. What if workers perceive that their greatest risk of injury is a fall, but your most common injury involves overextension? This perception is not necessarily inaccurate. It could be that your workplace has many areas where falls are possible and have occurred in the past, but workers' focus on that problem has reduced that type of injury. However, if workers are focused on falls and not on overexertion, their efforts may not be as effective. The reason for measuring perceptions should not simply be to measure but also to find opportunities to correct and improve.

The best single set of data to ground perception measurements is a customized Pareto analysis of site or organizational accident data. This is a profile of your safety issues as they are currently manifesting themselves; and comparing this reality to the measured perceptions of the workforce can be quite revealing about the accuracy or inaccuracy of perceptions. Ultimately, most improvement involves changing the realities of the organization, but some improvements can be accomplished simply by correcting inaccurate perceptions of the way things are. Effective actions cannot be based on misinformation or inaccurate perceptions.

Measuring Aspects of Common Practice

The most direct impact on safety performance is the way people do their jobs. Some definitions of culture focus on "the way we do things around here" or "what people do when you are not looking." Culture is rightly more the reason for, or the influence on, these actions, but the actions themselves are the most visible product of culture. Common practice is a set of human behaviors, and behaviors are, by definition, observable. It seems logical that at least one effective way to measure common practice is to observe it.

Of course, observation, without some limitations, is subjectively evaluated and random. The way to make observation more representative and objective is to follow the guidelines for good sampling. A representative sample of common workplace practice is still quite a complex achievement unless we simplify it by focusing on one aspect at a time. The STEPS process defines how to identify and prioritize safety issues. Action plans focus on solving one problem or on addressing one issue at a time. If these plans involve what people should do, these actions can be observed.

The term "observation" has been tainted by behavior-based safety (BBS) processes and has come to mean more of a peer coaching than a pure observation. Common practice can be measured by pure observation without peer-to-peer feedback or interpersonal contact, if desired. These skills can be taught quickly and easily, and focusing on only one aspect makes such observation both easy and quick. It can be combined with other processes such as safety audits or safety blitzes. It can also be done by BBS observers if there is already such a process in operation.

The important factor in such a measurement is the trending over time. An action plan might focus on a particular procedure or a precaution. As the workforce begins to focus on this practice, the occurrence of the desired action should increase. The observations can simply count the number of times the desired activity is observed and the number of times it is not being performed. This number can be calculated as a percentage of desired action and tracked for trends. If a good sampling strategy is used, then the trends will better reflect actual practice and not just differences in the sample size and distribution.

The trends need to be communicated as well as measured. The trending information can be influential in maintaining focus on the targeted improvement and in motivating further improvement. When done effectively, this type of measurement is highly effective as an indicator of personal and group performance toward a goal.

One of the reasons safety improvement is often difficult is that the measurements of safety performance are too far removed from the activities that produce them. When members of the culture can see more clearly and timely how their actions impact progress toward a goal, it is effective and motivational. The next step in metrics is to show how these targeted activities impact the lagging indicators we traditionally view as the ultimate measurement of safety.

Measuring Impact on Lagging Indicators

All the measurements suggested in this section are process indicators and ultimately can and should be compared with lagging indicators. As we form and complete more and more action plans, change perceptions, and change trends in common practice, how do these impact accident rates, severity, and costs over time?

There has been a lot of criticism of lagging indicators in the safety community and a questioning of whether or not they are an accurate measurement of safety performance. Many safety programs and safety professionals are evaluated strictly by these indicators. Although they are not perfect indicators, they do have a value, especially when combined with other metrics into more of a balanced scorecard for safety. If we measure our activities to improve safety and understand how they impact the culture's perception of safety, then measure how changes in perceptions impact changes in common practice, then measure how changes in common practice impact changes in lagging indicators, we begin to have a deeper understanding of how we can create excellent performance in safety. We also more clearly see how creating the proper clarity of purpose, climate, chemistry and capabilities enables us to drive these improvements. For an example of a balanced scorecard for safety, see Figure 5.2.

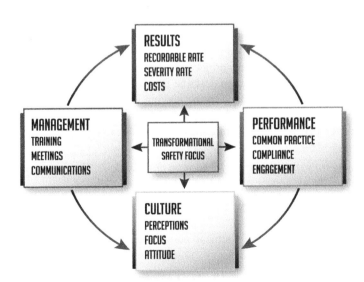

Figure 5.2   A sample balanced scorecard for safety excellence.

Another important outcome of performing a Pareto analysis of your organization's accident data will be to determine if your lagging indicators are measured, detailed, and documented well. Almost all organizations keep the required documentation to determine accident frequency and severity rates, but not all the data that drives these calculations have the same richness of detail. Accident reports need to have the right variables to help to better understand what happened and how it can be prevented in the future. If such data are computerized, it often makes the information more searchable and easier to analyze. For a list of recommended variables on accident reports, see Appendix D.

Moving Toward a Balanced Scorecard for Safety

As you move from measuring only lagging indicators in safety to measuring other aspects, keep in mind where this journey of adding new metrics might eventually lead. We believe that current trends in safety are leading to a "balanced scorecard" approach to safety measurement.

The original idea of a balanced scorecard began in strategic management. Managers used lagging indicators almost exclusively to manage businesses for many years. The history of these metrics dates back to the Industrial Revolution when the gentry began to fund businesses and developed metrics to hold managers accountable for their running of the business. These metrics proved valuable for accountability, but they did not help the managers to manage better. Most businesses knew how much money came to them as income, how much they spent in expenses, and how much became profit.

In the early 1990s, a business professor, Robert S. Kaplan, and a consultant, David P. Norton, began to collaborate based on a study of how they believed organizations would measure performance in the future. They developed a theory that allowed for a four-part set of metrics that centered around a business strategy and measured the steps needed to fulfill that strategy at critical stages. They found that this model of measurement helped to develop better strategies and better techniques for managing the business and for projecting changes needed in the future. It is well beyond the scope of this chapter to describe in full all the benefits or facets of this approach, but they basically added to the lagging indicators three other categories of measurement to complete the four-part model. They asked these four key questions and developed ways to measure their success in each category:

1. *Lagging indicators basically focused on the question.* "To succeed financially, how should we appear to our shareholders?"
2. *Customer indicators.* "To achieve our vision, how should we appear to our customers?"
3. *Learning and growth.* "To achieve our vision, how will we sustain our ability to change and improve?"
4. *Internal business process.* "To satisfy our shareholders and customers, what business processes must we excel at?"

Each of these impacts the others and combines to produce the overall performance results. They each also reach back to the vision and strategy for validation and direction. When done effectively, they have helped strengthen vision and developed effective strategies and metrics to ensure success in accomplishing the vision.

So how could this model apply to safety? We, like the strategic managers, already have lagging indicators. They are useful to hold safety managers accountable, but they do not necessarily tell them how to manage safety better. So what if we ask the same questions and design metrics to answer them? Here are the questions as I would propose them for safety:

1. *Lagging indicators.* To achieve our vision, how should our safety lagging indicators look? Obviously, we want accident and severity rates to decrease or disappear. We want to control the costs of accidents as well. We already have metrics for these and can examine them to see if they can be further improved.

2. *Customer indicators.* Who are our customers in safety? It is our employees including ourselves! So how should our safety efforts look like to them? Should they be included and/or represented in designing them? Should they be constantly informed on the focus and status of such efforts?

3. *Learning and growth.* How do we learn and grow better safety performance? STEPS is designed to be an excellent answer to that question. Will our customers appreciate that we are constantly learning and growing, and can we measure that progress and share it with them?

4. *Internal business processes.* What processes do we need to improve safety? Meetings, training, supervision, and STEPS. How will we measure them to ensure that we are working our processes and that our processes are producing the desired results?

If you utilize the measurements suggested in this section, you will already have some or all you need to begin to formulate a balanced scorecard for safety:

1. *Lagging indicators.* Recordable rates, severity rates, trends in rates, and costs.

2. *Customer indicators.* Perceptions, grounded perceptions, and common practice.

3. *Learning and growth.* Progression through STEPS.

4. *Internal business processes.* The development of the clarity, climate, chemistry, and capabilities (also the continuation of traditional safety efforts and programs that were in place before beginning the STEPS process).

The development of such a set of safety metrics can greatly aid in the improvement of safety performance. You can see how each effort produces results in each of the four metric sets and thus produces the desired results. As you utilize the multiple metric approach, you will begin to develop a feel and then more exact measurements of how one set of metrics impacts the others. Ultimately, these algorithms can help even more to direct efforts to improve safety. When we begin to understand how much effort produces how much change in perceptions and common practice and how much those changes impact processes and results, we develop a deeper and more profound knowledge of how safety performance works and how to improve it.

Case Study: A division vice president (VP) for a client company sent us an e-mail describing how new metrics had changed his job. He said that for years, he felt that he was driving his site managers to chase the cycle of avoiding failure described earlier. He said that all the organizational metrics were failure metrics, and he felt that he had to use punishment and threats to accomplish his goals of continuously failing less in safety. He admitted being skeptical when we first challenged his organization to define what safety success looked like and how to measure that. But he stated that his job now is encouraging success rather than punishing failure. He feels like a coach rather than a cop and has better professional relationships with his site managers. He reported that the communications between them was much more positive as were the results as measured by lagging indicators.

Application of STEP 5.8: We find that safety metrics are often strongly entrenched into organizations and that it is difficult to make significant changes to them quickly. However, most organizations are looking for leading indicators to supplement their lagging indicators and are open to adding new metrics slowly and checking them out before making more changes. The journey to better safety metrics needs to be taken slowly and methodically in most organizations, and this STEP can be the beginning of that journey. Add the mapping through the STEPS and ongoing measurements of perceptions and accident-data Paretos and let them become cultural before moving on. If your organization is more open to rapid change in safety metrics, research them or get professional assistance to customize a balanced scorecard approach for your organization.

STEP 5.9   TRUST (THE BONDING AGENT)

These are the elements of a molecule of excellent safety culture, but we have not yet discussed what holds them together. *Trust* is the bonding agent of this molecule, and it is critically important to the success of Safety Culture Excellence. Have you ever worked for someone you did not trust? Did you come to work everyday dedicated and motivated to do your best? Have you ever worked for someone who did not trust you? Did you feel they gave you the support you needed to be your best? Did they give you creative input in your own work and help you to have a sense of ownership? Translate these individual feelings to a whole organization that lacks trust and see what devastating effects it can have. Alternatively, imagine an organization where trust permeates everything they do. Can you also imagine what it feels like to work there and the pride you could take in producing excellence in everything you do? Can you imagine the sense of teamwork and cooperation that is possible in such an organization?

Trust levels can be measured to some degree in your perception survey and interviews, but realize that low trust levels can also skew the data from both of these. Too much scrutiny while administering the surveys or an overreaction to results can impact trust levels. Most people in touch with the organization and its people have a "feel" for what the trust levels are. Trust is one aspect of culture that is difficult to measure discretely. But an inexact measure of trust is better than no measure at all. Trust levels are often reflected in willingness to cooperate or participate in voluntary activities. It is almost always present when people are willing to go "above and beyond" in their personal efforts for leaders or for the organization. Some organizations poll their SET members regularly to see if they sense trust levels improving or degrading. A sense of the trust levels is better than no metric, and organizations that care about people want a sense of where these levels are.

The most effective way we regularly see to destroy trust is through the misuse of discipline. Organizations that overuse discipline or administer it unfairly can drive trust levels down very quickly. Even when discipline problems are corrected, the trust levels seldom rise quickly. Lessons that are learned painfully are seldom forgotten. Painful memories are significant barriers to trust between the giver and the receiver of the pain. Even the friends or associates of the pain giver are suspect to most victims. Pavlov in his famous dog experiments was not able to lure the dogs back into the rooms where they were punished.

Case Study: In our safety culture assessments, we regularly ask if discipline for safety is used often and if it is administered fairly. The responses to this question are usually one of our most predictive indicators of what the trust levels in the organization are. During these questions at a site, we got an overwhelming and almost unanimous response that discipline was not fair. When we probed further, we found that almost everyone in every group told the same story of two workers who did a similar thing, and one was severely punished, and the other was not punished at all. After hearing the story three times, we started asking when the incident happened. No one was exactly sure when, but they thought that it was around 1975 or 1976. This was an extreme case but far from the only such experience we have had. Even the perceived unfair use of discipline can create distrust that lasts a long time and is difficult to rebuild.

Application of STEP 5.9: Someone once compared trust with a 5-gal bucket that must be filled with an eyedropper but can be kicked over and spilled all at once. It takes a significant amount of time to build trust but only a nanosecond to destroy it. If trust levels in your organization are high, protect them and foster their growth. If trust levels are low, begin to build them and take great care not to destroy your progress with a careless word or reaction. Trust is the result of a history of reliability. If you say what you are going to do and regularly do what you say, trust begins to build. The moment you do not do what you say, the trust can be weakened or totally destroyed. New trust is fragile and only grows stronger with time and demonstration of reliability. Do not try to rush the building of trust but patiently begin to build the foundation. Consider the level of trust as a bank account you have with each person or group in your organization. Every time you meet or communicate, you either make a deposit or make a withdrawal from that trust account. Keep track of your deposits and withdrawals and know your balance.

CONTROL

> *All science is concerned with the relationship of cause and effect. Each scientific discovery increases man's ability to predict the consequences of his actions and thus his ability to control future events.*
>
> —*Lawrence J. Peters*

Milestones 1–5 were designed to help your organization develop the capabilities to support and encourage the growth of excellence in your safety culture. Milestone 6 is designed to help you identify safety challenges, prioritize them, and address them one at a time. This milestone and the next, final milestone have STEPS, but you are quickly approaching the point where you will define the STEPS. You will define your own safety-improvement targets and start addressing them. There is little in safety more satisfying than actually overcoming a safety challenge and moving on to another. These STEPS are truly the basic units of improvement. Once your culture has mastered the capability to improve, it has developed the most basic capacity for excellence.

This methodology is not simple a problem-solving protocol. Its goal is not to provide a reactive band-aid solution to a pressing problem. It is also not a method of placing blame. It is adapted to addressing needs and opportunities for safety improvement in a very proactive way. The ultimate goal of an excellent safety culture is to get past the reactive stage of safety and continue to improve without accidents stimulating all significant actions.

Goals: To enhance the culture's ability to identify risks

To enhance the culture's ability to prioritize risks for maximum effectiveness

To teach the Safety Excellence Team (SET) to develop action plans to address risks

To help the SET to communicate ongoing action plans

To help the SET to communicate progress toward success of action plans

STEPS to Safety Culture Excellence^{SM}, First Edition. Terry L. Mathis and Shawn M. Galloway.
© 2013 John Wiley & Sons, Inc. Published 2013 by John Wiley & Sons, Inc.

To align behavior-based safety (BBS) (if the site has a process) with STEPS

To align safety motivational strategies with the success of action plans

Methods: Accident-investigation data analysis

Pareto analysis

Developing action plans

Communicating action plan details and progress metrics

STEPS: 6.1 Targeting Safety Improvement

6.2 Taking a Safety-Improvement STEP

6.3 Converting BBS to STEPS

6.4 Motivation

STEP 6.1 TARGETING SAFETY IMPROVEMENT

The first STEP of controlling your safety performance is to target specific safety improvements. Targeting involves two parts:

1. Identifying opportunities to improve
2. Prioritizing those opportunities

Identifying Opportunities to Improve

The most critical skill for a "can-do" culture is how they "do." The first part of doing is deciding what to do. This necessitates having a methodology for targeting specific improvements. If you have already developed strong methods for solving safety problems, you can certainly utilize them here. An opportunity to further enhance them may be in this area of targeting.

The key to successful targeting of safety improvement involves working on the right things (the highest impact items), the right-sized projects (not too much at a time), and working on them the right way (utilizing effective problem-solving techniques). Effectively improving safety is a matter of identifying and addressing risks. Risks, once identified, can be addressed either by improving workplace conditions or by improving common practice or by a combination of the two. Even problems that are complex and involve multiple solutions may be best solved one STEP at a time.

To identify and prioritize improvement opportunities, your SET should analyze the data available from your existing safety efforts and programs. There are many forms or analysis to determine risks or hazards in the work place. If your organization has performed or hired consultants to perform a workplace risk analysis, you can use the results to help you with this step. If you have never performed a formal analysis, however, we suggest that you simply follow the guidelines below to begin to determine how to focus your efforts. At most sites with mature safety programs, the major risks have been identified and addressed either through conditional or through behavioral risk controls. The next challenge is to find the missing or weak links in this safety net and the opportunities to improve. The place to start this effort is with your accident reports from the past 3 years or so. In these reports, we need to look for commonalities in the following areas:

- Types of accident: trip/fall, fall from height, struck by, and so on
- Nature of injury: burns, cuts, abrasions, and broken bones
- Type of employee injured: regular, temporary, contractor, and so on
- Location of accidents
- Time of the day
- Time on the shift when the accident occurred
- Tenure of employee injured
- Jobs being done by workers when accidents happen
- Regular jobs versus cross-trained jobs versus temporary assignments
- Tasks being performed
- Tenure of injured employee: with the organization, on this particular job.

Use your organizational accident and/or near-miss data to perform an analysis to determine variables involved most in accidents at your site. The idea is to focus your efforts where they will produce the quickest and most significant results. This type of analysis does not ignore the fact that there can be risks that have not yet turned into injuries or that some risks may have the potential to produce more severe injuries than they have so far. This can always be true. However, if you address the issues that have manifested themselves already in accidents and then move on to the more hidden and subtle risks, you will most often have the best results for your efforts.

If your accident-investigation reports lack the detail and richness of data to allow you to do some or all of this type of analysis, it is time to consider revising your forms and methods of accident investigation. There is a list of variables that we recommend you include in your accident-investigation reports in Appendix D. The ultimate goal of accident investigation is to learn lessons that will aid in the efforts to prevent future accidents. If the current process fails to do this in any respect, consider revising it.

Once you have a good understanding of the details of your accidents, it is time to turn your attention to prioritizing the various strategies for addressing them. A tool to help with this is Pareto analysis.

Prioritizing the Opportunities

Eventually, you will want to address all the risks that cause or can cause accidents at your site, but how do you prioritize which risks should be address first? The primary tool to help with such prioritization is Pareto analysis, which is based on the Pareto principle. Named after Italian mathematician Vilfredo Pareto, the Pareto principle was proposed by American business-management consultant Joseph M. Juran, who was a colleague of W. Edward Deming. Deming and Juran pioneered statistical process control (SPC) and many other concepts still used in manufacturing quality control and effectiveness.

In contrast to problem-solving methods or root–cause analysis, this form of Pareto analysis is focused on achieving the targeted goal rather than determining causation or fixing the blame. It may be necessary to analyze secondary causes and influences on the target to determine the best way to improve. While this analysis seeks to make a dynamic difference, it also simultaneously prioritizes potential solutions in order of their impact on the issue. The result of such an analysis is a targeted improvement with high potential impact.

Pareto's original concept involved land ownership in Italy. He observed that 80% of Italy's land was owned by 20% of the population. He studied other countries and found similar distribution of ownership. Based on this 80/20 concept, he started to apply this kind of analysis to economic areas such as wealth and income distribution. Following his model, the UN Development Program Report of 1992 showed that the richest 20% of the world's population is controlling 82.7% of the world's income. After World War II, Juran began to apply this thinking to other business issues and concluded that, in most instances, about 20% of the effort produced about 80% of the results. He applied this principle to prioritizing certain efforts over others to produce faster results. He concluded that, theoretically, if you did only the most effective 20% of a job, you would produce 80% of the results that would be produced by doing the whole job.

In safety, this Pareto Principle can help to prioritize safety efforts. If we can identify and prioritize the few safety-improvement efforts that will produce the most results, we can do those first. Doing so will produce faster and more visible results that will help the culture become motivated and involved in reaching toward safety excellence. It will also allow the minimum use of resources, both capital and human effort, to produce the maximum efficiency in reducing accidents. The use of Pareto analysis can also help to ensure that a lot of effort is not spent first on activities that produce little if any improvement. It is important not to create the perception that the process is not working simply because it is focusing on the wrong priorities. False starts make it harder to succeed the next time.

A Pareto analysis should be performed on site or organizational accident data. Depending on the amount of such data, it is wise to use several years of data to complete the first analysis and determine the first priorities. If there have been substantial changes in the workplace (new equipment, location moves, redesign, etc.) or if there have significant changes in the workforce (downsizing, reorganizations, mergers, etc.), it is wise to use only the data accumulated since the last significant change, since earlier data may no longer be representative of the current issues.

Accident data can be analyzed and compared with the most common conditions and common practices that can impact accidents. Depending on the amount of detail in accident-investigation data and whether the data are entered into a computerized database application, the analysis can take more or less time to complete. It is important to determine both conditions and practices that could prevent accidents, not just those determined to be significant or root causes. It is also important to distinguish between practices that are mandatory (laws, guidelines, rules, procedures, etc.) and those that are discretionary (ergonomics, body position, communication, etc.). You should also keep track of how many of your accidents have no obvious prevention strategy if there are any. These data may help to improve accident-investigation techniques or point to future opportunities for improvement or problem areas to address. A worksheet and instructions for your Pareto analysis are in Appendix B.

From your Pareto analysis, you should choose your first TARGET for safety improvement. If it is a condition, you can develop a plan to address it and inform everyone of the plan. If it is a behavior, your action plan should involve publicizing and encouraging the taking of a precaution to reduce the probability of an accident. This targeting of safety improvements is a product of transformational thinking. Driven by Pareto analysis, transformational thinking analyzes what one thing if changed or done differently would make the greatest difference in safety? In contrast to kaizen that targets small, gradual improvement, transformational thinking targets the change with the greatest impact.

However, the highest scoring item on the Pareto analysis may not be the best first choice if it is too complex or difficult. The goal is significant, but rapid change. Complex or difficult issues can delay results and may best be put further down the priority list so that rapid and visible progress can begin. The other consideration is the ability of the SET to solve safety problems. As they gain experience and have successes, their ability will grow. Therefore, it makes sense to address more easily solvable problems first to hone their abilities before asking them to take on the more complex and challenging issues. When you decide on a safety-improvement target, you should take it to the next STEP.

STEP 6.2   TAKING A SAFETY-IMPROVEMENT STEP

Once you have identified a safety-improvement target, you need to turn your targeted improvement into an action plan. An action plan should be completed on the form in Appendix E or a similar form. (Many organizations have an action item form and follow-up reporting system on their computer software for safety or other functions, and, if you have them, you should utilize them for this project as well. This can avoid redundant paperwork and keep the action plans in a familiar and usable format.)

An action plan has several basic parts: *define* the goal, *who* will do *what* by *when*, and how we will *follow up*. The target identified in the risk analysis and prioritized in the Pareto analysis is obviously the goal or what is to be accomplished. The other parts are basically assignments made to individuals with various targeted completion dates and follow-up strategies. The idea is to mobilize the right members of the safety culture to accomplish the goal.

It is not enough to develop and accomplish action plans. You must also communicate these activities and results. This communication is aimed not just at informing but also at involving everyone in the culture in the targeted improvement.

Even if everyone does not have a specific assignment in the action plan, everyone is part of the culture and should be informed and aware. A team needs players, coaches, and fans. These roles can change as new STEPS are addressed, so everyone needs to keep track and learn how improvements are identified, prioritized, and accomplished.

As an action plan is put into place, communication should immediately follow. The entire safety culture should know what is targeted and who is assigned to do what. If everyone is aware, they can look for opportunities to render aid or to offer suggestions where needed. It is of great importance that the improvement efforts belong to everyone. Keeping information from people about what is going on is a sure way to alienate and separate part of the culture from the improvement efforts. The military has a term for confidential information that is not intended for everyone. They say that information is on a "need-to-know" basis. In STEPS, everyone has a need to know and should always be informed, both of targets and of progress toward goals.

Everyone should keep score and track the successes. Therefore, it is not enough just to inform people about targeted improvements but also to keep them appraised of progress toward their completion as well. One simple and effective metric for progress is the assignments from the action plan that are completed. This can be a raw number, but it is usually better if expressed as a percentage of the action plan completed. In some of the old charitable fundraisers, there was a picture of a thermometer or rain gauge, and it was colored in as a percentage of the goal as the contributions started to come in. Seeing progress toward the goal motivated further effort and usually speeded the completion of the drive. The same principle applies to safety improvement: a clearly defined goal and visible progress toward the goal motivate the members of the culture to complete the project and realize the improvement.

As discussed in the STEP for Communication, multimedia and repeated communications are often needed to effectively keep everyone informed. Make sure to utilize your best communication tools to convey success metrics. Nothing else will be as motivational to your process as this.

STEP 6.3   CONVERTING BBS TO STEPS

There are a lot of different things these days that are called BBS, but if you have a behavioral checklist, a behavioral observation process, and/or some way of utilizing observation data, these can be used for STEPS. Some organizations have healthy and well-functioning BBS processes and simply want to keep them independent but have them work harmoniously with STEPS. Others, for various reasons, want to modify or restart their BBS processes to become completely integrated with STEPS. Either of these can be accomplished as outlined here:

If you would like to maintain the integrity of your BBS process but make it work synergistically with STEPS, here are the ways to make them compatible and complimentary:

1. Keep your steering team/committee independent but have them meet with the SET periodically and coordinate their efforts.

2. Keep your site behavioral checklist, but focus observations on the STEP Targets and use observers to take the message about the targets to the workers and measure progress of practice/precaution target goals. Add the current target to the checklist or simply put an additional blank for it so you can gather data.

3. Share the percent safe on the targeted precaution with the SET so that they can measure success and know when to move on to another target.

4. When the target is a condition, have observers publicize the target during observations and gather any suggestions or comments from workers to help improve the condition.

If you want to use STEPS to restart or redesign your BBS process more completely, consider these suggestions:

1. Make the SET the new steering team for BBS or have the worker members of the SET as a subteam or subcommittee to steer the BBS process. If you have an existing steering team for BBS that you want to maintain, make them a subcommittee of the SET and facilitate meetings between them.

2. Do not have a checklist except the current improvement target. As you move on to new targets, keep the old ones on the checklist to maintain focus and make sure that there is no backsliding. State the target as a precaution and mark either safe (if workers are observed taking the precaution) or concern (if workers are observed not taking the precaution).

3. Have observers focus workers on the targeted improvement and do quick observations of only the one item. Give positive reinforcement to workers taking the precaution and simply express concern and try to find out the reason if workers are not taking the precaution. Do not have observers confront the workers observed or try to force change on the spot. Simply express concern and ask why. Bring the percent safe and the "why" data back to the SET.

4. Also utilize the observers to communicate to workers about progress toward improvement target goals and successes as they do their observations.

5. Crunch the observation data for the SET and have them utilize it to measure progress, declare success, and decide when to move on to the next improvement target.

6. Publicize the observers as another way for workers to send suggestions and comments to the SET.

Sites with BBS processes of various types have successfully utilized them in the STEPS process. Decide if your existing BBS process is healthy and functioning and should be utilized to help STEPS, or if it has become dysfunctional or stagnant and could be reenergized by a more complete makeover toward the STEPS methodology.

Lean BBS® was developed by ProAct Safety® to be compatible with STEPS as well as organizations using Lean Manufacturing.

STEP 6.4   MOTIVATION

An excellent safety culture is one that is motivated toward continuous improvement. New research into what really motivates people is quickly making traditional ideas of motivation obsolete. If your organization has an existing safety motivational program or strategy in place, we ask you to consider improving it but not to suddenly stop it. The opposite of motivation is demotivation, and that is exactly what occurs when incentives or rewards are suddenly taken away. Any changes in your motivational strategies should be gradual and well communicated. If, on the other hand, you do not have a safety motivation program in place, you can chose to add one if you think that it could add value. If you do so, you can utilize the new technology and avoid the problems that have plagued traditional programs.

First, if you have an existing system of rewards, incentives, or recognition, you should carefully examine what goal the incentives are attached to. In other words, exactly what is being motivated? If the motivation is attached to lagging indicators (like giving a bonus for completing a period of time without an injury), you definitely need to retarget. Regulatory agencies are beginning to discourage and even punish programs that pay workers to "not have accidents" on the grounds that they can potentially discourage workers from reporting accidents. There are instances in which workers have gone to great lengths to avoid reporting an accident that would have lost a bonus for themselves and their coworkers. Remember that depending on the way you structure incentives and rewards, you can inadvertently reward a behavior you did not intend to. It is important to constantly ask, "Are we rewarding the right thing?" and, "How else might someone get the reward besides the way we intend?"

If your motivational strategies are targeting activities rather than results, we suggest that you align them with the STEPS activities. The goal should be to motivate improvement, not just activity. The STEPS activities are designed to help the organization accomplish improvement, so motivating people to follow the STEPS can be a precursor to actually rewarding the results accomplished. Existing motivators can be gradually converted to reinforce the STEPS activities, then the process indicators of the action plans, and ultimately the changes in common practice that will produce the improved performance.

The goal is to motivate improvement, and recent research indicates that one of the strongest motivators is "visible progress toward goals." A critical part of the motivational strategy is to measure and communicate progress toward goals. The more visible the progress becomes, the more motivational it will be to the safety culture. When you target a specific improvement and the workers see the goal becoming a reality, they feel that they are part of a success story. This is where the motivational strategy merges with the communication strategies. Utilizing the right media to communicate progress toward goals, or developing new or multiple media for this purpose will contribute to motivation.

The success of the motivational strategy can be measured, in part, through the measuring of perceptions. The perception survey can determine if motivational strategies are working and to what extent they impact a percentage of the workforce. The tracking of these perceptions can help to determine the effectiveness of the communications and help to fine-tune it. Never forget that communication is a two-part process. Just because a message is being sent does not necessarily mean that it is being received. The perception survey can measure the reception part of the communication formula.

STEPS can simplify the use of safety incentives and rewards by dividing all safety efforts into basic units of accomplishment. These units contain goals for improvement and excellent opportunities to reward accomplishment without the traps and contingencies of many traditional motivational strategies. The safety culture is directly impacted by the STEPS process and internalizes the basic methodology of setting and accomplishing goals as a group. Rewarding and celebrating these accomplishments can ultimately become the safety incentive program. What better to incentivize than improvement?

CONTINUOUS IMPROVEMENT

There is no finish line.
—Nike Corporation Motto

There is a joke about continuous improvement that it was ok once but has gone on too long now. Unfortunately, some of the continuous-improvement efforts become a joke by losing either the "continuous" or the "improvement" part along the way. STEPS is designed to ensure that continuous improvement becomes sustainable and imbeds itself into the safety culture. The stability of using the same method to solve all safety problems also begins to dissolve the "flavor-of-the-month" mentality that pervades so much of today's work on safety. You can work on different aspects of safety and rotate your focus while utilizing the same basic improvement process. Involving as many people in the improvements as possible makes the effort cultural in its very nature. STEPS can be used to address organizational support, behaviors, conditions, or any other target for safety improvement without the need to implement a whole new program or process. Taking the "constant change" feel out of safety methodology can make it less de-motivating and more effective.

Continuous improvement is a necessary element of excellence. Although you have completed many STEPS to build characteristics and capabilities into your culture to support excellent safety performance, they will need to be revisited periodically. The safety strategy can become ineffective, the clarity you created can become clouded, the climate and chemistry can change, and the amount of control over safety performance will be impacted by these changes. As you continue to take STEPS to improve your safety performance, you will also need to take occasional STEPS to rebuild the support structures that make them possible. Do not view these as failures but simply as preventative maintenance. All systems are subject to entropy and will require attention. Even a ship that is sailing well toward its destination will need to plug an occasional leak and mop the decks.

Remember how many times we have said that excellence is a journey and not a destination. Even though you have taken almost all the STEPS toward it, this is and will continue to be true. There is no resting on your laurels or holding your

STEPS to Safety Culture Excellence[SM], First Edition. Terry L. Mathis and Shawn M. Galloway.
© 2013 John Wiley & Sons, Inc. Published 2013 by John Wiley & Sons, Inc.

ground. You must constantly raise the bar of your expectations and never settle for "good enough" along the way. Even as you move through your action plans and see your safety challenges getting smaller and fewer, you cannot quit taking the next STEP. Some say that the ultimate STEP will take you to zero accidents, but even that status has to be maintained. Getting to zero is a challenge; staying at zero is often even more challenging. Do not think of these as the last STEPS but as the last kind of STEPS you will need to learn since they will be needed regularly on the rest of your journey to Safety Culture Excellence.

The first, defined STEPS helped build your safety culture's ability to reach higher and meet greater challenges. You shaped the clarity, climate, and chemistry of your culture to enable you to develop the ultimate capabilities to identify, prioritize, and address your safety challenges. This last challenge will never be conquered or even caged. You will be working on it as long as your organization continues to fulfill its mission.

Goals:	To set clear expectations about continuous improvement, STEP by STEP
	To align levels of the organization in support activities
	To establish a rotation plan for the Safety Excellence Team (SET)
	To establish a new-employee orientation to STEPS
	To identify professional development activities for the SET and cultural leaders
	To provide guidelines for execution of the safety strategy
	To recommend ongoing assessments to identify improvement opportunities
Methods:	Evaluation
	Action plans to address problems or opportunities to improve
	Attending events or obtaining professional development materials
STEPS:	7.1 Ongoing Safety-Improvement STEPS
	7.2 The FILM for a Cultural Snapshot
	7.3 Multilevel Support
	7.4 Succession Plan for SET
	7.5 Onboarding: New-Employee Orientation to STEPS
	7.6 Professional Development
	7.7 Reassessment

STEP 7.1 ONGOING SAFETY-IMPROVEMENT STEPS

The safety-improvement STEPS you learned to take will be one part of your ongoing pursuit of Safety Culture Excellence. No matter how good you become, you will always be looking at your Pareto analysis and be targeting a new improvement. As you do so, continuously look for ways to improve that process. Can you do better ongoing Pareto analysis? Can you get more people involved and better focused to help you make the targeted improvement? Can you communicate your progress and successes more effectively? Can you celebrate your successes and recognize those involved to better motivate the workforce?

As you continue in this process, you will move from one target to another. As you solve your greatest safety issues, the amount of accident data in that category will begin to decrease. Often, before the data disappears (as the problem is solved), it will begin to lose its statistical significance. This will make it difficult or impossible to detect trends or to determine strategies that will be effective in addressing the remaining issues. For this reason, you should plan to move further upstream in the accident process to gather data so that you can continue, as long as possible, to have enough information to be statistically significant and drive effective safety improvement.

If you do not already have one, you should implement a "near-miss reporting process" to encourage workers to report injury-free events. Even as you begin to run out of accidents that produce injuries, you will still have near misses. These are often directly proportional to the risks of more serious accidents. Near-miss data can be predictive of where the next accident could potentially happen. As accident data diminishes, you should rely more and more on near-miss data to determine your improvement targets.

Accident reduction efforts often impact the highest severity levels first and have a later impact on lower severity events. Even though this is not always the case, it is almost always true that a significant number of near misses persist even when injury-causing accidents have not happened for a period of time. It is good to prepare for this eventuality before it happens. So, as you start to address your injuries, begin to encourage near-miss reporting so that there is no gap between the types of data to drive continuous improvement.

As the culture reaches greater levels of excellence in safety performance, even the near-miss data can start to become insufficient. If you already have a behavioral observation process, you can utilize it to replace the disappearing near-miss data. As we mentioned in the last milestone, you might want to revise your observation process to better fit the need of the STEPS methodology. If you do not have a behavioral observation process, you can establish one in a fashion much easier than traditional behavior-based safety (BBS) processes. Many sites utilize their safety reps to do quick observations on the one improvement target as described in the BBS modification process.

So you should plan to progress from accident data to near-miss data to observation data to develop your ongoing safety improvement targets in your journey to Safety Culture Excellence. Preplanning can be the key to success.

STEP 7.2   THE FILM FOR A CULTURAL SNAPSHOT

This STEP is designed specifically for SET members but is often shared with the leaders, managers, and supervisors. The reason others find it helpful is because, like

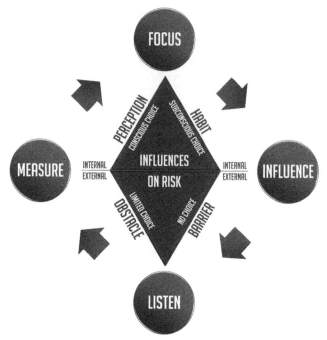

Figure 7.1   The FILM for a cultural snapshot.

many other models in this book, it is a performance-improvement model and not limited strictly to safety performance.

At this point in your progression through the STEPS, some SET members begin to wonder exactly what their job is going to be on an ongoing basis. They have been very engaged in a journey through the STEPS, and that journey seems to be over. Do they simply decide on a series of improvement targets and design STEPS to accomplish them? Exactly how will they monitor the safety culture to make sure it is growing in excellence? The answer to that question is that they will begin to take a series of snapshots of the culture to both see what it looks like and see how it is changing. These are not literal photographs, but they constitute a good look at the culture and what influences it. The process is described in Figure 7.1. The parts of the snapshot should prompt the SET members to ask and answer questions about each part. The term FILM is an acronym for the parts of the model.

"F" is for FOCUS: An excellent safety culture is focused on an improvement target. That does not mean that they are blind to other safety issues, but while they see the big picture, they are focused on the current STEP to improve. This means that anyone in the organization you ask should be able to name what the current focus is. They should know, at least approximately, when the STEP started and what, if any, progress has been made. If the organization is not focused or if the communication is missing areas or shifts, the SET can create action plans to address the issues. The lack of focus almost always creates the lack of improvement. One of the most critical capabilities needed for Safety Culture Excellence is the capability of creating focus that in turn creates progress. We like to use an acronym for FOCUS that illustrates what it is intended to accomplish: Forming One Common Understanding of Safety. In other words, this means getting everyone on the same page, getting all the arrows pointing in the same direction, getting the whole organization working on the same targeted improvement in safety, and taking the same STEP together. The SET should constantly be asking themselves and each other, "Are we focused?"

"I" is for *influence*: People do things for a reason. We call those reasons influences. If you do not change the influences, you might not have a permanent change in what people do. Do you know what influences the current performance in safety and how the influence or influences could be changed to support better performance? Many safety initiatives seek to change behavior through confrontation or peer pressure rather than seeking to deeply understand the influences that are reinforcing the behavior. The danger of ignoring influences is that they continue to work against your forced behavioral changes. The best way to understand why people do something a certain way is to ask them. If coaching becomes both the management style and the interpersonal communication style of the culture, then expressing concern and asking why become a cultural norm. The whole idea of influence is understanding and working with the "why" rather than the "what" and "how." The best way to improve performance is to target an improvement and to align the influences to reinforce the improved performance.

"L" is for *listen*: The SET should steer the STEPS process, but they should constantly have their proverbial "ear to the ground." The late Steven Covey admonished us to form the habit of understanding before we seek to be understood. This means listening before you speak. SET members should be constantly asking and listening to people. The SET can lead the culture toward excellence, but the people *are* the culture. We have helped many organizations improve their safety performance simply using the information we found by asking their own people how to improve. Listening actively and often is a key to Safety Culture Excellence.

"M" is for *measure*: It is hard to improve anything that is not being measured. Measurement is the way to know if something is improving or not, and improvement is a primary tool for motivation. The SET should constantly be exploring how to measure or better measure the targeted improvements. When they design a safety-improvement STEP, how will the improvement be measured and how will they know that the STEP is complete and that it is time to move on to the next STEP? We do not have discrete ways to measure every aspect of human performance, but we can learn and explore. Look how much improvement has been made in how judges rate human performance in the Olympics over the last decades. It is not perfect, but it is greatly improved. Such metrics and the progression of them should be on the minds of SET members regularly as they move toward a more excellent safety culture.

In the center of the FILM process is a diamond, which is the model we want to suggest to the SET to classify and address influences. When you see a member of your safety culture doing something and you want to understand what might be influencing it, this is the model to use. It divides the influences into four reasons which fall into two categories.

- Category 1: Influences within the control of the worker

  Perceptions. Workers may take a risk if they do not perceive it to be a risk or if they calculate it to be an acceptable or manageable risk. Remember that perceptions can be either accurate or inaccurate and that they are usually based on personal experience that differs from person to person. Perceptions can be changed through better or more accurate information.

  Habits. Workers develop patterns of behavior as they do tasks multiple times. After enough repetitions, workers can do the task without consciously thinking about it. It becomes automatic. Habits can be changed with spaced reminders that are regular and long enough to change the pattern and not fall back into the old pattern of behavior.

- Category 2: Influences not easily or at all in the control of the worker

  Obstacles. These are influences that make it difficult, awkward, or more time consuming to perform a task safely. Common obstacles involve the location of tools and equipment or personal protective equipment (PPE) in relation to where they are needed. Changing performance influenced by obstacles involves either removing the obstacle, redesigning work to avoid it, or convincing workers that the inconvenience is worthwhile to improve safety.

  Barriers. These are influences that make it impossible for the worker to take a precaution or to do a task safely. Common barriers include not having the right tool for the job, not having anything to tie off fall protection to, or broken seat belt buckles. Changing performance limited by barriers involves removing or working around the barriers.

The SET should view their ongoing responsibilities as designing safety-improvement STEPS and continuously improving the culture. The ongoing STEPS of continuous improvement and the current status of the safety culture can be evaluated regularly through this model of FILM and the reasons for risks. This should become the mental model for understanding where the culture is and what influences will need to change to move to the next levels of excellence.

STEP 7.3 MULTILEVEL SUPPORT

Safety Culture Excellence requires support from each level of the organization (site management, middle management, engineering or technical, supervisors, union representatives, workforce groups, etc.) Ideally, all or most of these groups will be represented on the SET. Each group's safety roles, responsibilities, and results (SRRRs) can and will likely be different. If sufficient support is well defined and ongoing, there is no need for further action. If needs change, problems arise, or lack of support is perceived or indicated in ongoing perception surveys, the SET should take action to assess and redefine the support needed to maintain the STEPS process. As improvement opportunities are found, action plans can be developed to improve support just as they can to improve safety performance.

A popular and often effective method to improve support in higher levels in the organization is to simply document the problem and to ask those involved to develop their own action plan to improve. Obviously, if individuals from that group are resistant or unresponsive, such an approach will not work. But most of the problems of upper level support are more a matter of multiple priorities than resistance. If they are resistant, interventions by groups like the SET that are outside the level where the problem exists may not be effective either.

The most common issues of lack of support from lower levels tend to be a lack of information or unclear expectations. These can often be addressed easily and in a timely manner either through meeting with the groups involved and sharing information and clarifying expectations or through supervisors, area managers, or BBS observers. Often, a member of the SET is from the area of the problem and can handle the matter by meeting with their work team.

The ultimate goal is for each level to rely on the ongoing perception surveys as well as introspection to evaluate their support roles and performance. Any feedback from the SET or other groups involved should also be taken into account. The goal is not to be adequate, but to be excellent and always improving. Each level should strive to be a champion of the STEPS effort and be viewed by the SET and facilitator(s) as valuable assets and allies.

STEP 7.4   SUCCESSION PLAN FOR SET

When you recruit the initial members of your SET, you should ask them to serve for longer or for indefinite terms. However, you want to give others in your organization the opportunity to serve on this team if there are those who desire to. The number of volunteers may be limited at first, but it should grow over time based on the success of the efforts. As workers see the process working and are kept in touch through the communication strategies, some of them will begin to envision themselves in the roles of leading the process. It is not necessary to give everyone an immediate opportunity to become a member of the team; in fact, allowing the membership to rotate more slowly can actually make membership more desirable if managed well. But keeping everyone waiting too long can backfire and damage the effort.

Another consideration is what we refer to as "critical mass." This is the idea that the SET has more expertise on STEPS than any other members of the culture and that their expertise constitutes a critical mass of knowledge and skills necessary to keep the process alive and progressing. If you replace too many of the SET members at once, you seriously diminish the critical mass of knowledge and endanger the process. The most successful rotation strategies make some new opportunities periodically without seriously diminishing the overall critical mass of expertise. For example, a SET with 8–12 members might rotate two members every 6–12 months, thus bringing in "new blood" but keeping enough old hands to make sure that the team knows what it is doing. New members can be brought "up to speed" usually in 6 months or less so that they are operational when the next rotation occurs.

Many organizations begin to recruit new potential SET members at 12–18 months into the process and actually begin the rotation at 24 months after the implementation. There are no magic numbers, but these approximate time tables seem to work for most projects. Reasons to vary from these timetables can be based on the site's specific makeup including tenure, turnover, stability of the organizational structure, management stability, and changing safety issues. The specific needs of the site should take precedence over any formula. The better the process fits the site and culture, the greater the chance of success.

It is wise at some sites to anticipate the possibility that there will be little or no desire on the part of employees to become part of the team. Sites with no history of participatory safety programs or with strained relations between unions and management lead the lists of sites with these challenges. When recruiting the original SET members, it is good to plan for such a contingency and to ask them to be willing to serve for a longer time if needed. At some sites that have no issues with getting new volunteers, it can still make sense to have minimal turnover of SETs due to their high level of performance and excellent teamwork. The performance of the SET needs to be carefully weighed with the needs for further involvement opportunities to find the right balance of SET rotation.

STEP 7.5   ONBOARDING: NEW-EMPLOYEE ORIENTATION TO STEPS

As you implement STEPS, you will eventually reach every employee in your organization. But, if you have future turnover, you will be bringing new employees on board who are not familiar with the process and do not know their part in creating an excellent safety culture. For this reason, you should include a basic overview of STEPS in your new-employee orientation to ensure that new folks know the process and get on board quickly. Some sites supplement the initial training by teaming new employees up with more experienced employees in a coaching or mentoring mode. This pairing can also speed the induction of new employees into STEPS and get their questions answered on the job. Sites with BBS processes working with STEPS often assign an observer to either mentor or frequently observe new employees until they become familiar with their new jobs.

In addition to sharing details of how the STEPS process works, it is important to show new employees where you are on the journey (which STEPS are completed and which are currently focused) and share the results you have had thus far. In other words, do not just explain the program, explain the benefits and make sure to answer the what's in it for me? (WIIFM) question. The initial marketing program of STEPS to the employees can be utilized here. The goal is not to inform, but to sell. We want buy-in from the new employees as well, and the quicker we can accomplish that after they start work, the stronger we make the safety culture and the more we control the potential dilution of the culture. You need to control the ratio of workers participating versus those not participating because they are still learning or unsold.

Sites have taken several approaches to assigning the new-employee orientation. Most of the sites have the safety department or safety manager design and deliver the orientation. Larger sites sometimes have a training department or human resources trainers who handle it. However, a growing number of sites are asking the SET to handle the orientation. Some SET members do this personally while others assign these duties to employees who are not team members. Sites who have assigned this aspect of new-employee orientation to workers claim that it brings the new workers into the culture more quickly and completely than when it is handled by managers or supervisors.

STEP 7.6   PROFESSIONAL DEVELOPMENT

Working on the initial STEPS can be self-explanatory, can keep the SET and facilitator busy, and can be self-motivating as you see progress toward goals. However, at some point, the leaders of the process will start to run out of new ideas and begin to feel like they are simply "going through the motions" if you do not plan for further learning, or professional development. Almost everyone involved in special projects or disciplines receives some type of continuing education or association with others working on the same issues. The further knowledge and sharing of ideas and experiences can help the process to continue to improve and to remain motivated. For these reasons, we recommend that toward the end of the first year of STEPS (or sooner if you are already in the Capabilities steps) that you plan for further professional development of your SET and related sponsors, facilitators, single points of accountability (SPAs), and so on.

Professional development can be accomplished in several different ways:

- *Published materials.* STEPS leaders can receive articles, new books, or white papers on safety culture and share them with each other. Many teams prefer assigning one team member to read the material and report to the rest of the team rather than giving each member reading materials. This approach tends to minimize reproduction costs and lessen the reading burden.

- *Internet.* STEPS leaders can download and share podcasts and participate in webinars or videoconferences or virtual conferences. Some SETs make the sharing of such materials a regular part of their meetings and discuss new ideas in light of existing issues or unsolved problems.

- *Site visits or conference calls.* Members of STEPS leadership or SET members can visit other sites in your organization or other organizations and share ideas from the visit with the others involved. This can be accomplished via actual trips or over conference calls. Actual visits are valuable for viewing processes, meeting in person, and viewing materials. Calls can minimize costs while involving more people and potentially utilizing more input and ideas.

- *Conferences.* Representatives of STEPS leadership can attend conferences and share the information on their return with the remaining leaders. Participation can rotate so that more people can share the experiences of exposure to others working on safety culture. Such attendance can become a reward for service in the process. Your organization or site can also host its own STEPS conference and may opt to invite outside experts (either from other sites or organizations or from professionals) to speak or lead workshop sessions.

The goal of professional development is to glean new ideas. These can include new approaches to process issues, problem-solving techniques, or motivational strategies. New ideas come from exposure to new information or people engaged in similar activities. To ensure that this happens, many organizations ask those attending events or getting information to briefly summarize new ideas and present them to fellow leaders and/or team members and to the facilitator. Formalizing the thoughts on paper or computer sets the expectations that reinforce the goals as well as helping to more effectively remember and communicate the new ideas.

STEP 7.7   REASSESSMENT

In addition to developing action plans to address your safety challenges, you need to periodically audit your culture on these first STEPS and see if you have maintained the status that will allow for excellent performance in safety. If any STEP is found lacking, you can develop an action plan to address it. It is common for excellent safety cultures to find that they have several new supervisors who are not effective safety coaches, that they have a communication medium that no one uses, or that they have lost focus in a department with high turnover. Action plans to enhance coaching skills, workshops to develop new communication media, or group meetings to reestablish safety focus can be part of action plans with assignments to SET members (or others) and planned completion and follow-up.

We recommend that you use the STEPS audit list (Appendix F) as well your ongoing perception surveys to determine your status on each primary STEP on an annual basis. Be careful not to turn this into a fault-finding mission or a witch hunt. The goal is not to become critical of your own performance, but to objectively determine where improvement is needed. Some SETs appoint members to become the audit team, and some sites also assign members of management to facilitate the process by providing resources, access, and advice. The findings are provided back to the SET to develop any needed action plans, and again, facilitators can aid with those.

CONCLUSION

Now that you have mentally gone through the series of STEPS and milestones, you should be ready to plan your organization's real journey. It helps to have the entire roadmap in your mind before you begin. It also helps to fully appreciate what will be required as you prepare to inform others and to help them make the commitment to pursue Safety Culture Excellence. It is not an easy journey, but it is incredibly worthwhile. Watching your organization grow in its ability to focus and build its capacities to accomplish and sustain excellence is deeply rewarding.

The order in which we have arranged the STEPS has come from a lot of experience. If you are planning to visit each step, please do so in order. As you move to a new STEP, please assess your current status in that area of performance. Skip over any STEPS that you have already addressed and simply review them to see if you find ideas to further improve. If you are planning to implement only a few ideas from this text, we recommend you do so in the same sequence they appear here. We sincerely believe that this advice will bring you the greatest success with a minimum of difficulty.

Remember that what we have provided you in this book is not a complete formula; it is more of a framework from which to make strategic decisions. Organizations are so unique that no formula for improvement is universally applicable. The closest to a universal truth is the fact that the more you customize this process to your organization, the better it will work. Also, as you assess where your current safety performance is weak and where it needs help, do not forget to look for your strengths and build upon them.

Keep the mindset of the gardener growing a plant. You already have a safety culture, and if you plant the seeds of excellence in it and mind the climate and chemistry, you will see the capabilities grow. Excellence is not necessarily perfection, but it is more a matter of personal best. Let your culture become excellent in its own unique way.

As you pursue the altruistic goals, do not forget that good safety is also good business. If we want safety to be integrated into company strategy, we need to integrate strategic thinking into safety. Only with sound strategic direction can we ensure that our activities will continuously add value. We sincerely hope that this book is valuable to you, but remember, tools are only as effective as the individual who use them. The real work is up to the readers.

STEPS to Safety Culture Excellence[SM], First Edition. Terry L. Mathis and Shawn M. Galloway.
© 2013 John Wiley & Sons, Inc. Published 2013 by John Wiley & Sons, Inc.

BIBLIOGRAPHY

Agnew, J., & Snyder, G. (2008). *Removing Obstacles to Safety: A Behavior-Based Approach* (2nd ed.). Atlanta, GA: Performance Management Publications.

Anderson, G. M., & Lorber, R. L. (2006). *Safety 24/7: Building an Incident-Free Culture.* Lafayette, LA: Results in Learning.

Dekker, S. (2008). *Just Culture: Balancing Safety and Accountability.* Farnham, Surrey: Ashgate.

Florczak, C. M. (2002). *Maximizing Profitability with Safety Culture Development.* Oxford: Butterworth-Heinemann.

Frost, P. J., Moore, L. F., Louis, M. R., Lundberg, C. C., and Martin, J. (1991). *Reframing Organizational Culture.* Newbury Park, CA: Sage.

Krause, T. R. (1996). *The Behavior-Based Safety Process: Managing Involvement for an Injury-Free Culture* (2nd ed.). Hoboken, NJ: Wiley.

Krause, T. R., & Hidley, J. (2008). *Taking the Lead in Patient Safety: How Healthcare Leaders Influence Behavior and Create Culture* (1st ed.). Hoboken, NJ: Wiley.

Lawrie, M., Parker, D., & Hudson, P. (2006, March). Investigating employee perceptions of a framework of safety culture maturity. *Safety Science*, 259–276.

McSween, T. E. (2003). *The Values-Based Safety Process: Improving Your Safety Culture with Behavior-Based Safety* (2nd ed.). New York: Wiley-Interscience.

Meyers, R. J., & Thomas, R. A. (2005). *180 in 180 Days: 7 Principles that Will Transform Your Safety Culture.* Baton Rouge, LA: Firefly Publishers.

Montana, P. J., & Charnov, B. H. (2008). *Management* (4th ed.). New York: Barron's Educational Series.

National Safety Council. (1999). In *Safety Culture and Effective Safety Management* (G. Swartz, Ed.). Chicago: National Safety Council.

Nelson, E. J. (2005). *The Pathway to a Zero Injury Safety Culture.* Farmington, NM: Nelson Consulting, Inc.

Parker, D., Lawrie, M., & Hudson, P. (2006, July). A framework for understanding the development of organisational safety culture. *Safety Science*, 551–562.

Pink, D. H. (2011). *Drive: The Surprising Truth about What Motivates Us* (Reprint ed.). New York: Riverhead Trade.

Roughton, J. E., & Mercurio, J. J. (2002). *Developing an Effective Safety Culture: A Leadership Approach* (1st ed.). Oxford: Butterworth-Heinemann.

Skinner, B. F. (1938). *The Behavior of Organisms: An Experimental Analysis.* New York: Appleton-Century-Crofts.

STEPS to Safety Culture Excellence^{SM}, First Edition. Terry L. Mathis and Shawn M. Galloway.
© 2013 John Wiley & Sons, Inc. Published 2013 by John Wiley & Sons, Inc.

Smetzer, J., & Navarra, M. B. (2007, January). Measuring change: A key component of building a culture of safety. *Nursing Economics*, 25(1), 49–51.

Smith, S. (2008). *Cultivating a Culture of Safety in Healthcare: A Systematic Approach to Root Cause Analysis*. Cambridge: QIG Publishing.

Spitzer, D. R. (1995). *SuperMotivation: A Blueprint for Energizing Your Organization from Top to Bottom*. New York: AMACOM.

Spitzer, D. R. (2007). *Transforming Performance Measurement: Rethinking the Way We Measure and Drive Organizational Success*. New York: AMACOM.

Stewart, B. (2002, September 30). Behavioral approach one tool in "total safety culture." *Westchester County Business Journal*.

Yiannas, F. (2008). *Food Safety Culture: Creating a Behavior-Based Food Safety Management System* (1st ed.). New York: Springer.

APPENDIX A

EXPLANATION OF TERMS

- *Behavior-based safety (BBS).* Originally a term used to describe the efforts of early behavioral scientists to apply behaviorism to safety. Later commercialized and adopted by many safety programs that tried to address worker behavior. The mainstream efforts in BBS almost all include several features:
  - ○ identifying and addressing specific behaviors believed to be critical to safety;
  - ○ observations in the workplace to see if these behaviors are happening or not;
  - ○ some form of feedback to workers on their behaviors; and
  - ○ the use of the observation data to try to more proactively manage safety outcomes.
- *Lean BBS.* A safety process utilizing traditional BBS, Lean Manufacturing, and STEPS technology designed to be implemented with a minimal use of internal and external resources and to produce quicker and more sustainable results than traditional BBS. The main differences between Lean BBS and traditional BBS are as follows:
  - ○ Use of existing infrastructure such as safety committees
  - ○ Shorter, more-focused checklists or single behavioral improvement targets
  - ○ Fewer, more-focused observations utilizing performance coaching methodology
  - ○ Better, more-integrated utilization of observation data
- *OSHA's voluntary protection plan (VPP).* An outreach program designed to foster cooperation between OSHA and organizations that choose to participate. It was aimed at increasing worker participation in safety efforts and called for a series of committees in which workers could discuss and address various aspects of safety efforts. A significant number of organizations utilized this program to foster a more "open-door" relationship with OSHA as well as to give employees a more active role in safety.
- *Total recordable incident rate (TRIR).* A ratio of accidental injuries to the number of hours worked. It was originally designed to standardize a metric for safety performance in a way that would allow comparisons of sites with differing numbers of workers and hours worked. The ratio is calculated by

STEPS to Safety Culture Excellence^{SM}, First Edition. Terry L. Mathis and Shawn M. Galloway.
© 2013 John Wiley & Sons, Inc. Published 2013 by John Wiley & Sons, Inc.

dividing the number of injuries that qualify as "recordable" by OSHA standards by 200,000 hours worked. The idea is that this number of hours approximates 100 workers working full time for a year (50 weeks per year at 40 hours per week × 100 workers). The problem with TRIR is that it has become the most recognized metric for safety, and this has driven many organizations to try to reactively manage their safety programs or otherwise manipulate this number to make them look like they are progressing or benchmarking well in their industry. We view it as a valid, but limited metric that needs to be supplemented with other measurements to truly measure and manage overall safety performance.

- *Lost time accidents and rates.* OSHA defines "lost time" as the time during which a worker cannot perform their normal duties due to work-related injury or illness. The calculation begins on the day after the accident or illness on which the worker cannot report and continues until the worker resumes normal duties. Many organizations have tried to focus on their most severe injuries in hopes of lowering the total impact of injuries. This approach has met with limited to no success in many industries due to the difficulty in accurately predicting the severity of injuries in accident types. The risks or injury type that results in the most severe accidents can vary widely from year to year, and picking one year's severe injuries to focus efforts on does not guarantee that next year's injuries will be less severe.

APPENDIX *B*

PARETO ANALYSIS WORKSHEET AND INSTRUCTIONS

Vilfredo Pareto was a multitalented man who had careers ranging from engineer to college professor. His greatest contribution to this methodology was a form of analysis he designed originally for economics. It is a way to quantify the potential impact of causes to overall impact. He first used it to prove that 80% of the land in Italy belonged to 20% of the families. We use it here to decide what safety focus would have the greatest impact on reducing accidents based on historical data. This approach has proven to be highly effective over the years.

Performing this analysis is relatively simple. You read an accident report and ask, "What could have prevented or reduced the severity of this accident?" You mark your answers in the categories on the following chart, total them, and calculate a percentage of the total number of accidents that this category of safety focus could have impacted. The use of this type of analysis can help focus safety efforts on the most impactful improvement strategies. Note: If you categorize the same accident under more than one category, your percent safes can add up to more than 100%.

Category	2 Years Ago	1 Year Ago	This Year	Total	%	Rank
Ergonomics						
Communication						
Body position						
Tool and equipment issues						
Material handling						
Personal protective equipment						
Rules and procedures						
Conditions and engineering						
No solution found						
Totals						

STEPS to Safety Culture Excellence^SM, First Edition. Terry L. Mathis and Shawn M. Galloway.
© 2013 John Wiley & Sons, Inc. Published 2013 by John Wiley & Sons, Inc.

As you begin to utilize Pareto analysis, you can either utilize or develop more precaution-specific lists to further narrow your focus of STEPS targets. There are lists available for housekeeping issues, workplace design issues, rules and procedures, and discretional behaviors (usually from BBS materials). To begin, categorize your issues and focus on targets to improve the category that would most impact accidents. When you have sufficiently improved or eliminated that category, celebrate your success and focus on the next most impactful category.

SELECTING EFFECTIVE MEMBERS FOR SAFETY EXCELLENCE TEAMS

A team is only as effective as its members. Selecting the right people for your Safety Excellence Team is a good start toward success. Use the following guidelines as you consider employees as potential team members.

The first consideration for team membership is to select a representative from each major area, department, or shift at the site (i.e., production, maintenance, and shipping). The team will make decisions for the rest of the workers so it is important for each major area to have a representative on the team.

Characteristics of effective team members:

- *Willing participant.* Good team members cannot be coerced into participation. It is not necessary that each team member volunteer, but he or she must be willing to participate when invited. However, many good team members have expressed a level of skepticism at first. Being skeptical is common and should not automatically eliminate a worker from consideration.

- *Respected worker.* Team members must be taken seriously by both managers and peers. Be careful not to mistake popularity for respect.

- *Experienced.* Novices seldom have adequate knowledge of the site, the people, or the safety issues to be effective team members. Do not necessarily select all "old timers," but avoid anyone who has not been on the job long enough to really know the ropes.

- *Good safety role model.* Team members do not have to be accident free, but they need to be serious about safety. Often, employees who have been injured develop serious attitudes about safety and have great credibility when they speak about safety to other employees.

- *Open-minded.* Team members need to be able to learn and manage new ways of doing things. Some people are overly opposed to anything new or different. Such people can seriously hold back progress if they are put in a position of leadership.

STEPS to Safety Culture Excellence[SM], First Edition. Terry L. Mathis and Shawn M. Galloway.
© 2013 John Wiley & Sons, Inc. Published 2013 by John Wiley & Sons, Inc.

- *Collaborative.* Team members need to be good team players. They need to work together and be able to come to consensus. Workers with reputations for stubbornness and not getting along with fellow workers should be eliminated from consideration.

- *Takes initiative.* Good team members should be self-starters. They should have demonstrated initiative on work projects or extra assignments.

- *Good communicator.* Team members should be able to effectively communicate with other workers and managers. They do not need to be professional speakers but should be able to effectively express ideas.

- *Good problem-solver.* Team members will be given data and expected to identify and solve safety problems. Workers with demonstrated ability to solve problems should be given strong consideration if they meet other criteria.

- *Personal organization.* Team members will need to manage multiple roles and priorities. Workers with the ability to get things done in an organized way often make good team members.

Selection Scoring Chart

Use the following chart to compare prospective team members. On a scale of 1–10 with 1 being the lowest and 10 being the highest, rate the prospective team member on each of the characteristics. Total the score by adding the row.

1	2	3	4	5	6	7	8	9	10

Lowest   Highest

Prospective Team Member	Willing Participant	Respected Worker	Experienced	Good Safety Role Model	Open-Minded	Collaborative	Takes Initiative	Good Communicator	Good Problem-Solver	Good Personal Organization	TOTAL SCORE

APPENDIX **D**

LIST OF VARIABLES RECOMMENDED FOR ACCIDENT INVESTIGATION REPORTS

Although we consider accident data to be a lagging indicator of safety performance, the reduction of accidents remains our goal. Accident data can help to focus STEPS on the most impactful targets and make our safety-improvement efforts more efficient and effective. When performing an accident investigation, there are some variables that can be reported and tabulated to help in this effort. These are in addition to the standard description of the event, the nature of injuries, and the responses performed. These are additional categories that can be analyzed and trended to better focus STEPS. The most common and useful ones include, but are not necessarily limited to, the following:

- Date:
- Time: (may include shift and time into shift)
- Consecutive days worked prior to the accident:
- Amount of overtime worked in past week and month:
- Location: (may include building or area, department, work station, etc.)
- Type of worker: (regular employee, contractor, temporary, summer intern, etc.)
- Tenure of worker: (in the organization and on the job being performed)
- Work or task being performed: (description but also was it the normal job, project work, pulled to different workstation to cover for absent worker, etc.)
- Status of facility: (normal operations, planned outage, unplanned outage, breakdown, emergency shutdown, etc.)
- Weather conditions: (details and comparison/contrast to normal)
- Special or unusual circumstances:
- Tools, equipment, and/or supplies involved:
- Others involved and/or injured:

STEPS to Safety Culture Excellence^{SM}, First Edition. Terry L. Mathis and Shawn M. Galloway.
© 2013 John Wiley & Sons, Inc. Published 2013 by John Wiley & Sons, Inc.

ACTION PLAN FORM

After performing your Pareto analysis, make a statement of your objective and the specific target of your improvement plan. Put in the Pareto percentage of accidents and the number of accidents in the targeted category. Fill out the following form on which you make assignments to accomplish the goals and track them to completion.

Action Plan Worksheet

Objective: _____

STEPS Target: _____ Date of Action Plan: _____

Targeted Completion Date: _____ Completion Date: _____

Data Tracking

Month	Beginning	Month 1	Month 2	Month 3	Month 4
Pareto %					
No. of accidents					

Action Item Assignment/Tracking

Person Assigned	Action to be Performed	Completion Date	Follow-Up

STEPS to Safety Culture Excellence[SM], First Edition. Terry L. Mathis and Shawn M. Galloway.
© 2013 John Wiley & Sons, Inc. Published 2013 by John Wiley & Sons, Inc.

STEPS AUDIT CHECKLIST

Each year, review the STEPS and see if you were already in good shape (Skipped) or needed to improve (Action Plan[s]). List the date your improvement plans were completed and what results indicated success.

STEP	Skipped	Action Plan(s)	Date Completed	Results
1.1 Purpose				
1.2 Core values				
1.3 Vision				
1.4 Long- and short-term goals				
1.5 Objectives				
1.6 Marketing				
1.7 Initiatives				
1.8 Safety Excellence Accountability System				
1.9 Identify and Enable Change Agents				
1.10 Measure/adjust				
1.11 Continuous improvement				
2.1 Evaluation of existing safety initiatives				
2.2 Perceptions				
2.3 Interviews				
2.4 Safety data analysis				
3.1 Structure				
3.2 SET strategy briefing				
3.3 SET clarity workshop				
3.4 Employee briefing(s)				
4.1 Commitment				
4.2 Caring				
4.3 Cooperation				
4.4 Coaching				
5.1 Passion				

(*Continued*)

STEPS to Safety Culture Excellence[SM], First Edition. Terry L. Mathis and Shawn M. Galloway.
© 2013 John Wiley & Sons, Inc. Published 2013 by John Wiley & Sons, Inc.

STEP	Skipped	Action Plan(s)	Date Completed	Results
5.2 Focus				
5.3 Expectations				
5.4 Proactive accountability				
5.5 Reinforcement				
5.6 Vulnerability				
5.7 Communication				
5.8 Measurement				
5.9 Trust (the bonding agent)				
6.1 Targeting safety improvement				
6.2 Taking a safety-improvement STEP				
6.3 Converting BBS to STEPS				
6.4 Motivation				
7.1 Ongoing safety-improvement STEPS				
7.2 The FILM for a cultural snapshot				
7.3 Multilevel support				
7.4 Succession plan for SET				
7.5 Onboarding: new employee orientation to STEPS				
7.6 Professional development				
7.7 Reassessment				

INDEX

STEPS to Safety Culture Excellence^{SM}, First Edition. Terry L. Mathis and Shawn M. Galloway.
© 2013 John Wiley & Sons, Inc. Published 2013 by John Wiley & Sons, Inc.